THE PERIODIC TABLE

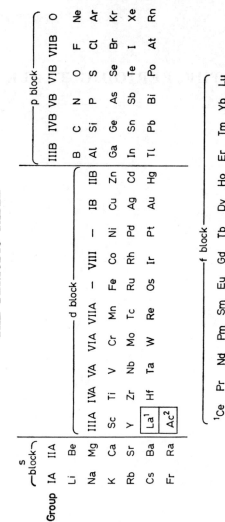

THE PERIODIC TABLE

D. G. COOPER, B.Sc., F.R.I.C.

*Head of the Science Department,
Birkenhead Technical College*

FOURTH EDITION

NEW YORK
PLENUM PRESS
LONDON
BUTTERWORTHS

Published in the U.S.A. by
PLENUM PRESS
a division of
PLENUM PUBLISHING CORPORATION
227 West 17th Street, New York, N.Y. 10011

First published by
Butterworth & Co. (Publishers) Ltd.

First Edition 1958
Second Edition 1960
Third Edition 1964
Fourth Edition 1968

©
Butterworth & Co. (Publishers) Ltd.
1968

Suggested U.D.C. No. 541·9

Library of Congress Catalog Card Number 68–56902

Made and printed in Great Britain
by William Clowes and Sons, Limited, London and Beccles

FOREWORD

The periodic system as developed by Mendeleeff and others in the latter half of the last century provided a systematic arrangement of the elements in their natural groups. The significance of this classification extended much further, as it permitted the properties of unknown elements and their undiscovered compounds to be predicted with some certainty. Later developments provided the theoretical basis for the periodic system and the point was soon reached when it was possible to understand fairly fully the valence properties of the elements and the nature of the chemical bond. The systematization of Inorganic Chemistry within the framework of the periodic system has formed the background to the progress of Inorganic Chemistry and has opened up the new fields of development which have become so important in recent years.

It is important that the modern student should appreciate and understand at an early stage the framework within which his subject is built rather than be overburdened with factual information, and in presenting a concise and readable account of the Periodic Table and its application Mr Cooper has done a valuable service to Inorganic Chemistry. The book is ideally suitable for Advanced Level teaching in the Grammar Schools and for First Year University and Higher National Certificate Students, and can be warmly recommended.

C. E. H. BAWN

The University of Liverpool

CONTENTS

	Page
FOREWORD	v
INTRODUCTION	ix
THE PERIODIC TABLE	1
Diagonal Relationships	11
Fajans' Rules and Ionic Potential	13
Position of Hydrogen	13
Occurrence of the Inert Pair	15

s-BLOCK ELEMENTS
   Group IA—The Alkali Metals . . . 17

s-BLOCK ELEMENTS
   Group IIA—The Alkaline Earth Metals . 21

p-BLOCK ELEMENTS
   Group IIIB—Boron Group . . . 24

p-BLOCK ELEMENTS
   Group IVB—Carbon Group . . . 28

p-BLOCK ELEMENTS
   Group VB—Nitrogen Group . . . 31

p-BLOCK ELEMENTS
   Group VIB—Oxygen Group . . . 37

p-BLOCK ELEMENTS
   Group VIIB—The Halogens . . . 43

p-BLOCK ELEMENTS
   Group O—The Noble Gases . . . 47

d-BLOCK ELEMENTS
   Group IIIA—Scandium Group . . . 50

CONTENTS

Page

d-BLOCK ELEMENTS
  GROUP IVA—TITANIUM GROUP . . . 52

d-BLOCK ELEMENTS
  GROUP VA—VANADIUM GROUP . . . 55

d-BLOCK ELEMENTS
  GROUP VIA—CHROMIUM GROUP . . . 59

d-BLOCK ELEMENTS
  GROUP VIIA—MANGANESE GROUP . . 62

d-BLOCK ELEMENTS
  GROUP VIII—IRON GROUP 66

d-BLOCK ELEMENTS
  GROUP IB—THE NOBLE METALS . . . 72

d-BLOCK ELEMENTS
  GROUP IIB—ZINC, CADMIUM AND MERCURY . 76

f-BLOCK ELEMENTS
  LANTHANONS AND ACTINONS . . . 80

ELECTROPOSITIVITY AND ELECTRO-NEGATIVITY 86

TRANSITION ELEMENTS . . . 91

OCCURRENCE OF THE ELEMENTS . . 97

RADIOACTIVITY AND NUCLEAR STABILITY 100

THE CHLORIDES OF SOME ELEMENTS . 103

TABLES OF PHYSICAL CONSTANTS . 104

BIBLIOGRAPHY 114

INDEX 115

INTRODUCTION

The classical development of Science has come through observations, theories to correlate these observations, and deductions from the theories capable of being tested. The general trend has been towards the logical simplicity of a single all-embracing theory, though even if we isolate one of the 'sciences' we cannot yet find that this stage has been reached.

Chemistry seems to many beginners to consist of a mass of unrelated facts, which make a big demand on the memory, and therefore act as a barrier to further progress. It is not until this barrier has been surmounted that the system behind the facts slowly becomes apparent, and the subject can be seen as a logical whole. Traditional methods of teaching the subject may to some extent be blamed for this belated appreciation. Even the convenient division of the subject into 'organic' and 'inorganic' has the drawback of delaying a consciousness of unity through the full application of atomic theory.

At an intermediate level, students frequently consider organic chemistry more systematic than inorganic; this is because systematic inorganic chemistry involves a well-developed knowledge of the Atomic Theory and its application to the Periodic Table. This, in turn, demands more of the student's understanding than does the integration of the facts of organic chemistry, and may result in the student regarding inorganic chemistry as sheer drudgery.

In an endeavour to help correct the balance between the branches, the present small work is offered, as a collection of essays on the Periodic Table explaining, and explained by, the Atomic Theory.

INTRODUCTION

Since this represents the logical background to inorganic chemistry, any facts set forth must be presented in as logical a way as possible, with the emphasis placed on their relevance to the general picture. No attempt is made to make the treatment comprehensive; it is a question of selection and deployment of relatively few facts.

The plan of the work starts with an account of the Periodic Table as a whole, and is followed by chapters on each group in turn; in accordance with the current practice, the groups are labelled 's block', 'p block', etc., but the older group numbers should be clear.

The later chapters deal with a few special topics which sometimes present difficulty to the student.

The third edition represented extensive revision, and the present edition has again been thoroughly revised and almost completely rearranged. An introduction to orbital theory has been provided to fit in with the form of the Periodic Table now used.

The book has apparently been found useful, and the author would welcome suggestions for further improvement.

The author would like to express his thanks to his colleagues at Birkenhead Technical College who have helped in the preparation of this book by reading the manuscript and making various helpful suggestions.

D. G. C.

THE PERIODIC TABLE

In the later sections the Periodic Table is dissected and examined group by group. It is first necessary to attempt a balanced appraisal of the whole Table, its structure, weaknesses and strength.

Mendeleeff was not the first to attempt a systematic classification of the elements, or to examine them in the order of their atomic weights. He was, however, the first to put forward as a periodic law, the gradation of properties observed when adopting this arrangement, and to employ it for prediction of the properties of elements for which he left spaces. In making predictions, he used the fact that similar elements came under each other, that is, fell in the same groups, while neighbours in his horizontal periods showed gradual changes. The form in which the Table is displayed has varied from time to time, and 3-dimensional models as well as charts have been used. One of the most useful forms of the Table, which has been modified by a number of workers, is given as a Frontispiece, and is based directly on the electronic structures of the elements.

Both physical and chemical properties show periodicity; for instance, density and melting point start at low values with the alkali metals and rise to maxima, then decrease again to very low values with the halogens. The atomic volume, which is obtained by dividing atomic weight by density, gives a curve which is of course almost the inverse of the density curve: the alkali metals have very high values, and the elements in the middle of each period have very low values, so that a repeating pattern is obtained with successive periods. This is the familiar Lothar Meyer atomic volume curve (*Figure 1*). Although atomic volume is a physical property, it has a profound effect on chemical properties; the size of the atom is one of the main factors considered by Fajans in his rules. By plotting the atomic volumes of each group [see *Figures 2* (*a*) and (*b*)] the relationships between the groups and subgroups become clearer.

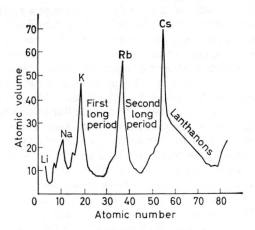

Figure 1. Lothar Meyer curve

As self-consistent figures of atomic and ionic sizes have now become available, it is usual to replace the concept of 'atomic volume' by that of atomic radius, or ionic radius, as appropriate. However, those conclusions which were based on atomic volume have not been invalidated. If allowance is made for the volume of a sphere, $v = \frac{4}{3}\pi r^3$, the figures are similar in relative order, whether we use atomic volume or atomic radius. In each case, the figures should not be regarded as absolute values, but from any one source figures have values which are reasonably correct relative to each other.

The graph (*Figure 3*) shows clearly the periodic variation in the size of atoms, and of the ions that can be formed from them; it will be noted that cations are

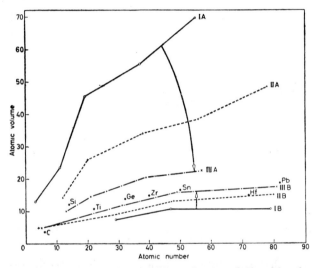

Figure 2 (a). Atomic number–volume relationship for Groups I to IV. Group IV points shown thus ·Sn

always smaller than their neutral atoms, while anions are always larger. Lothar Meyer was the first to notice this periodicity of size, as mentioned above, at about the time Mendeleeff was noting periodicity of other properties. His ideas were overshadowed by those of Mendeleeff, but size relationships are now of increasing importance in valency consideration and in crystal structure.

THE PERIODIC TABLE

The most obvious chemical property that varies through the periods is valency. In the first two short periods, the valency equals N, the group number (also equal to the number of electrons in the outer shell of the atom) or, for later groups, the alternative value of (8−N). This is because these elements react in such a

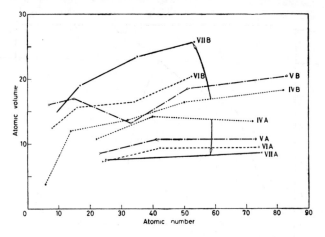

Figure 2 (b). Atomic number–volume relationship for Groups IV to VII

way that the atoms acquire the stable electron configuration of the nearest noble gas, an element which is non-valent* as it already has a very stable configuration. In the long periods, the valency pattern shows more variety, as only the first few and last few elements have single valencies equal to N or (8−N). The central elements in each long period belong to transition series, characterized by, among other properties, variable valency. These elements cannot readily attain a rare gas structure, except by coordina-

*But see p. 47.

THE PERIODIC TABLE

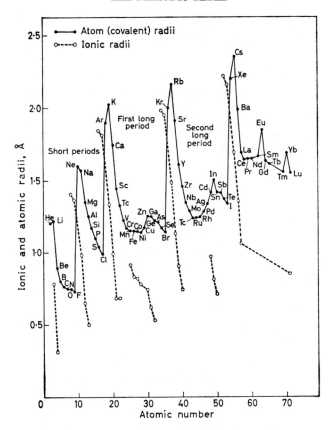

Figure 3. Atomic and ionic sizes

tion (see later), and the other possible structures do not differ very much in stability, so that chemical oxidation or reduction can readily bring about changes in the valency, which may rise to a maximum equal to the group number (see p. 93).

A further chemical property which varies through a period is the tendency to gain or lose electrons. This may be measured by the electron affinity and ionization potential, respectively. Elements in early groups, having a small number of electrons in their outer shell, readily lose these electrons and become positive ions—they are electropositive, or metallic. This makes their hydroxides basic, because of the ionization $MOH \rightarrow M^+ + OH^-$. Elements in later groups are progressively less electropositive, or more electronegative, until we have in Groups VI and VII families of elements capable of giving simple anions by acquiring two electrons or one electron per atom, respectively. These elements have acidic oxides, which give acids with water. Elements in the middle groups do not give simple ionic compounds as a rule, but form covalent compounds instead. Oxides of these elements may be amphoteric, or neutral, or feebly acidic.

While there exists this variation of electropositive character with group number, there is also a natural tendency for large atoms to lose electrons. In the case of both early and late groups, the heavier atoms are large, and therefore within these groups elements become more electropositive as atomic number increases. In the middle groups, however, and particularly among the transition elements, greater atomic weight does not mean any great increase in atomic size, so that elements in these groups do not necessarily become more electropositive with increasing atomic number (see p. 89).

The net result of the two effects is the 'diagonal relationship' of elements sometimes found, where one element resembles another in the next higher group and next later period.

The form of the Periodic Table is governed by the

atomic structures of the elements, and successive periods represent the addition of successive shells (see Frontispiece). The terms 's block', 'p block', etc., now in common use refer to the sub-shells which are being filled with electrons. The first shell cannot be divided into sub-shells, because it only contains *one* orbital, the *1s* orbital, spherical in shape and which can accommodate up to 2 electrons of opposite spin. Each orbital, in whatever sub-shell, can accommodate two electrons, as long as they are of opposite spin. In the second shell, there are s and p sub-shells, the first of which has one orbital (holding 2 electrons), and the second has three orbitals (6 electrons). The three p orbitals are elongated, and are in the three mutually-perpendicular planes. The third shell adds five orbitals, with room for ten d electrons, so that there are *3s*, *3p* and *3d* types of orbitals, while the fourth adds seven more orbitals which can hold 14 f electrons. The letters stand for sharp, principle, diffuse and fundamental, from names applied to spectroscopic lines which were thought to correspond with these particular electron-orbitals. The table summarizes the available orbitals in successive shells.

Shell	Orbitals				Total orbitals	Total electrons
	s	p	d	f		
1	1	—	—	—	1	2
2	1	3	—	—	4	8
3	1	3	5	—	9	18
4	1	3	5	7	16	32

The orbitals are usually represented by circles, as follows: (for the second shell) ○ ○○○, one s and three p orbitals. As orbitals are filled, arrows are placed in the circles in alternate directions, corresponding to opposite spin:

THE PERIODIC TABLE

No.		2s	2p	
3	Lithium	①	○ ○ ○	} 's' block elements
4	Beryllium	⑪	○ ○ ○	
5	Boron	⑪	① ○ ○	
6	Carbon	⑪	① ① ○	
7	Nitrogen	⑪	① ① ①	} 'p' block elements
8	Oxygen	⑪	⑪ ① ①	
9	Fluorine	⑪	⑪ ⑪ ①	
10	Neon	⑪	⑪ ⑪ ⑪	

It will be noticed that in carbon and nitrogen, there are two and three electrons respectively, in p orbitals but not paired. Pairing only starts when it is forced to—after each of the available orbitals has one electron in it; the number of electrons with unpaired spin is a maximum (Hund's maximum multiplicity rule).

The numbers of electrons which can be accommodated account for the numbers of elements in the periods; $2s + 6p = 8$, $2 + 6 + 10d = 18$, $2 + 6 + 10 + 14f = 32$. However, these maximum numbers are reached at different stages, and not in successive periods, where the totals are 2, 8, 8, 18, 18, 32. This 'delayed action' before the totals reach 18 or 32 is because the order in which orbitals are filled by electrons depends on the energy levels associated with the orbitals. The orbitals first filled are of lowest energy, and the filling takes place in the order of increasing energy-level. The process starts as one would perhaps expect, with *1s*, *2s*, *2p* orbitals first filled, then the *3s* and *3p* orbitals in elements 11–18, corresponding to elements 3–10 in the table. However, the next to be filled are *not* the *3d* orbitals but the *4s*, which happen to be of lower energy than the *3d*.

THE PERIODIC TABLE

The diagram (*Figure 4*) acts as a convenient memory-aid, for the arrows drawn from left to right cross out the orbitals in the order in which they are filled.

If all orbitals were filled, the total of elements would be 120, so further orbitals are not needed.

Shell 1   2   3   4   5   6   7   8
    s(2) s(2) s(2) s(2) s(2) s(2) s(2) s(2)
         p(6) p(6) p(6) p(6) p(6) p(6) —
                d(10) d(10) d(10) d(10) —
                        f(14) f(14) —

Figure 4.

			$3d$	$4s$	
19	Potassium	$(3s+3p$–filled)	○ ○ ○ ○ ○	①	's block'
20	Calcium	$(3s+3p$–filled)	○ ○ ○ ○ ○	ⓘ	elements
21	Scandium	$(3s+3p$–filled)	① ○ ○ ○ ○	ⓘ	'd block'
22	Titanium	$(3s+3p$–filled)	① ① ○ ○ ○	ⓘ	elements

The $3d$ orbitals are then progressively filled before the $4p$ orbitals. Elements 21–29, scandium to copper, have a (nearly) constant outer electron configuration, and are adding electrons in an inner shell, i.e. the third shell when there are already two electrons in the fourth shell. Such elements have long been known as transition elements, and are placed in the d block of elements. Transition elements usually possess certain properties which are not found to a great extent outside their ranks; these properties have been explained in terms of atomic structure. The common feature of atomic structure for the transition elements is that they have incomplete electronic shells other than the outermost shell (see Chapter on Transition Elements, p. 91).

The d block of elements finishes with the zinc group; however, if a definition such as that given above is used, these elements are not transitional in structure or character. Zinc itself has a completed $3d$ sub-shell, and also has the two s electrons in the fourth shell. Because of the complete $3d$ sub-shell, zinc does *not* show variable valency, does *not* have coloured ions, and does *not* show particular catalytic activity. It does, however, show a marked tendency to form complex ions. The inclusion of the zinc group in the d block is not illogical, but is parallel with the inclusion of the noble gases at the conclusion of the p block.

The s block and p block of this modern periodic table emphasize the vertical groups, and the elements in these blocks are normally considered in their 'families'. Within the d block there is less scope for vertical group comparisons, although in some cases this treatment is still a fruitful exercise; the shape of the block invites a horizontal comparison. Various physical properties can, for instance, be plotted on a graph, when it will be found that most of these properties show comparatively slow changes from one end of the block to the other.

The f block elements do not come into the scope of intermediate syllabuses, but are analogous to the d block elements. They are known as 'inner transition' elements, because the f orbitals do not fill until there are some electrons in the *two* outer shells, i.e. the first row occur when the $4f$ orbitals are filling, and there are $5s$, $5p$ and $6s$ electrons present. These elements are known as Lanthanons, and the final row ($5f$) are known as Actinons; these include transuranic elements, made synthetically. Some further details are given in the appropriate chapter.

FAJANS' RULES AND DIAGONAL RELATIONSHIPS IN THE PERIODIC TABLE

The reason for the well-known diagonal relationships found among some elements in the Periodic Table is simply the incidence of the two Fajans rules. These rules are concerned with the ease of formation of electrovalencies and covalencies, that is, in the case of a metal, with the electropositivity of the element.

The first rule says that covalencies are more likely to be formed as the number of electrons to be removed or donated increases, so that highly charged ions are rare or impossible. The removal (or donation) of a second electron must overcome the effect of charge due to the removal (or donation) of the first, and so on for each successive electron. The work required quickly becomes impossibly great for chemical forces as the number of electrons involved increases, and covalencies result instead.

The second rule states that electrovalencies are favoured by large size of cations or small size of anions. In a large atom the outer electrons are further from the attractive force of the positive nucleus, hence are more easily removed to give cations; while in small atoms the added electron to give an anion can approach more closely to the positive field and will therefore be more strongly held than in a large anion.

The operation of the first rule in passing from Group I to Group II is approximately balanced by the effect of the second in passing from the first period to the second. The three most obvious examples of similarity are the following:

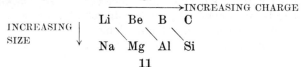

Lithium shows its similarity to magnesium in all those points which emphasize its difference from sodium. Examples are the relative instability of its nitrate and carbonate to heat; the sparing solubility of its fluoride, hydroxide, carbonate and phosphate; the high average degree of hydration of its salts; and its reactions with nitrogen and carbon.

Beryllium shows so close a resemblance to aluminium that it was originally thought to be trivalent. The two have a similar reaction to acids and alkalies; the oxides and hydroxides of both are amphoteric. Both metals form compounds which are mainly covalent, and often largely hydrolysed. Beryllium chloride is almost as effective as aluminium chloride in Friedel–Crafts reactions. The main differences can often be explained by the fact that, in complexes, aluminium can show a covalency maximum of 6, while for beryllium it is 4. The beryllium ion can only be hydrated to the extent of $4H_2O$, while for aluminium this figure is often exceeded.

Boron and silicon also show strong resemblances to each other. The elements are similar in inertness, high melting point, and the (slow) reaction with alkali rather than acid. The oxides are acidic and form glassy non-volatile solids, and borates and silicates also have parallel properties. The hydrides are similar in preparation, volatility, hydrolysis and some reactions. Borides and silicides have similar properties—they are infusible and unreactive. The halides of the two elements are covalent, volatile and easily hydrolysed.

Examples of diagonal resemblances can be found in other pairs of elements, and since the line separating metals and non-metals runs diagonally, there are naturally several metalloids along it. However, in these other cases the diagonal relationships are usually less strong than the family resemblances within a group, whereas in the case of beryllium there is probably a stronger resemblance to aluminium than to magnesium or zinc, and with boron the resemblance to silicon is stronger than to aluminium or gallium.

FAJANS' RULES AND IONIC POTENTIAL

We saw (p. 11) that electrovalency is favoured by small ionic charge and by large cationic size. Cartledge attempted to relate these two factors in one expression, which he called ionic potential. His expression was

$$\text{ionic potential, } \phi = \frac{z}{r}$$

where z is the charge on the cation, and r its radius.

By examining the results for several metals, he was able to show that if ϕ^i is less than 2·2, the oxide of the metal is basic. If ϕ^i is between 2·2 and 3·2, oxides are amphoteric, and if above 3·2, oxides are acidic.

THE POSITION OF HYDROGEN IN THE PERIODIC TABLE

There have from time to time been discussions about the position of hydrogen in the periodic table. It has been quite usual to show it in two positions, both above the alkali metals and above the halogens.

A closer examination shows that resemblances to either Group I or Group VII elements are somewhat superficial, as perhaps might be expected in view of the very great differences between the two groups.

Hydrogen shares with the alkali metals its univalency and ability to form cations; but unlike Na^+, etc., H^+ does not exist except in the solvated form,

$H_2O \rightarrow H^+$ or H_3O^+, and HCl is physically quite unlike NaCl*. The structure of the atoms appears temptingly similar—each has 1 electron in the outer shell. However, the hydrogen atom alone has an electron with no inner shells, which explains why H^+ has no separate existence (in solution) and why hydrogen is much less electropositive than lithium.

Again, hydrogen shares with the halogens the ability to form univalent anions and the atomicity of the covalent molecules of the elements. HCl, HBr, etc. are intermediate in physical properties between H_2 and the corresponding halogen molecules. However, H^- is incapable of existence in water, since H_2 is immediately formed, and there is no stabilization due to solvation; in any case, hydrogen has much less electron affinity than the halogens, so that H^- is much less common than halide ions. There is only superficial resemblance between OH^- and OCl^-. The atomic structure of hydrogen is one electron short of a rare gas structure. However, H^- has two electrons to one proton, a higher ratio than for any other anion, and clearly an unstable structure is the result.

In thermodynamic properties (ionization energy, electron affinity), hydrogen has most similarity to carbon and other Group IV elements. This is borne out in organic chemistry, for the C—H bond has less polarity than the bond between carbon and any other element. It may, of course, be pointed out that in hydrogen atoms the first shell is half full, as in carbon the second shell is half full. There is as much justification for showing hydrogen above Group IV, therefore, as there is for placing it above the alkali metals or the halogens, and no single place is satisfactory.

*Metals cations also are *normally* hydrated in solution.

OCCURRENCE OF THE INERT PAIR

The 'Inert Pair' is a pair of outer-shell electrons, not forming part of a complete octet, not being employed for valency purposes. Nitrogen has 5 electrons, and in its 3-covalent compounds (e.g. NH_3) there is a pair of electrons not being used. However, these *are* part of the octet and are referred to as an 'unshared pair'.

The inert pair effect is found mostly among some of the heavy elements, and the most obvious result is to produce a valency 2 less than the group valency. Thallium (Group III) can be trivalent but forms stable monovalent compounds. Tin and lead (IV) both show electrovalencies of 2 as well as 4, and divalency is predominant in the case of lead. Bismuth (V) is predominantly trivalent, and so on. The occurrence pattern is for the inert pair to be found in the later B subgroups, in the last period (IIB—Hg—unreactivity, effective valency 0) and then in earlier periods as the group number increases—roughly a triangle.

IIb	IIIb	IVb	Vb	VIb	VIIb
			P	S	Cl
Zn	Ga	Ge	As	Se	Br
Cd	In	Sn	Sb	Te	I
Hg	Tl	Pb	Bi	Po*	At*

Sulphur and chlorine are considered to show inert pair behaviour; in their case the pair of electrons is additional to the complete octet (the rather rare

4-covalent sulphur, 8 shared and 2 unshared; 3-covalent chlorine, as in ClF_3, 6 shared and 4 unshared).

The inert pair is thought to be a pair of s electrons; but it is difficult to see why the occurrence pattern is as shown. It would be perhaps reasonable to find this phenomenon among heavy elements generally, which tend to be less reactive, more 'noble'. Also, it is more common for atoms to lose electrons than to gain them, and easier for atoms to lose a small number of electrons than a large number; bismuth, at valency 3, is quite definitely metallic in nature.

Elements in A subgroups have electronic structures with a 'rare gas' core, while elements in B subgroups have not; the penultimate shell in the case of B subgroup elements contains 18 electrons. Although 18 electrons represents a stable configuration, it is not often found as an outer shell and is apparently rarely stable under these conditions. The bismuth atom, for instance, has the structure 2, 8, 18, 32, 18, 5. The Bi^{3+} ion is therefore: 2, 8, 18, 32, 18, 2. This represents a more symmetrical distribution of electrons than a structure having a 'bare' 18 electrons in the outer shell, and electronic arrangements often show a high degree of symmetry. The 2 electrons in the outer shell buttress the penultimate group of 18. A similar argument applies to all the other elements in B subgroups.

It is possible to explain the 'inert pair' effect in terms of ionization potentials for removing successive electrons, e.g. for bismuth the figures are:

$$8 \cdot 5, \ 16 \cdot 8, \ 25 \cdot 7, \ 45 \cdot 5, \ 56 \cdot 2 \ (eV).$$

The sharp jump from the 3rd to the 4th electron is obvious.

s-BLOCK ELEMENTS—GROUP IA: THE ALKALI METALS

Lithium	2.1
Sodium	2.8.1
Potassium	2.8.8.1
Rubidium	2.8.18.8.1
Caesium	2.8.18.18.8.1

The alkali metals are all soft, silvery-white metals which rapidly tarnish in air. They can readily be cut with a knife. Their melting points are low, decreasing steadily with increasing atomic number from lithium (180°C) to caesium (28°C). Their densities are also low, rising from a figure of about 0.53 g/cm^3 for lithium to 1.90 g/cm^3 for caesium. The atomic volumes are therefore high; in each case these elements have the highest atomic volumes in their respective periods.

Electrical conductivity is very high for all the alkali metals, and so also is their power to emit electrons under the impact of light. Both potassium and caesium have been used in the construction of photo-cells; the latter element is the best available of all metals for this purpose.

The great chemical reactivity of the metals is best shown by the increasing violence with which they react with water. They will in fact continue to react with ice even at very low temperatures. The hydroxides produced are in each case solids of high melting point, extremely stable to heat, but readily soluble in water to give strongly basic solutions:

$$MOH \rightarrow M^+ + OH^-.$$

The electropositive nature of these elements is shown by the properties of their hydroxides, and also by those of their salts. All the common salts are freely soluble in water, even the carbonates, and the solutions are very highly ionized (lithium salts are least ionized, and some can apparently become covalent). The salts are usually more stable to heat than are the corresponding salts of other metals. For instance, the carbonates do not decompose below their melting points, which are high, except for slight decomposition in the case of lithium carbonate. The bicarbonates of alkali metals are the only ones which are obtainable as solids stable at room temperature.

The chlorides (and other halides) have high melting points, and conduct electricity in the fused state. The usual method of preparing the elements is by electrolysis of the fused chlorides; it is necessary to bring down the melting point by the addition of other chlorides as impurities. Potassium, rubidium and caesium further show their strongly electropositive nature by the formation of per-halides, e.g. $K^+I_3^-$.

The metals react with hydrogen on warming to give solid, salt-like hydrides. These have high melting points, and on electrolysis at or near the melting point give hydrogen at the anode. The structure must therefore be M^+H^-.

The electropositive properties, and the great reactivity, of the alkali metals are of course explained by their atomic structure and large atomic radius. Only one electron must be removed to give the stable rare gas structure. In the case of lithium, the atom is already comparatively large, and as the atomic size increases with atomic number, so the single valency electron becomes progressively more easily removed (that is, the elements are progressively more electropositive). This is illustrated by the figures in the

Appendix, p. 104. Both monovalent charge and large atomic radius favour the formation of cations, $M \rightarrow M^+$, according to Fajans' rules.

The chemistry of the alkali metals is therefore that of their ions, as there is very little tendency for complex formation. This is due, on the one hand, to the great stability of the readily attainable ion structure, and on the other, to the instability of any readily attainable covalent structure.

On comparing the corresponding salts of the alkali metals, a decreasing tendency for hydration to occur as atomic number increases is seen. Nearly all lithium salts are hydrated, and many sodium salts are hydrated, or hygroscopic. Few potassium salts are either hydrated or hygroscopic (compare nitrates, chlorides, chromates and dichromates of sodium and potassium). No common rubidium or caesium salts are hydrated. That this is a property of the metal ions is shown by comparing their mobilities, which increase in the order Li^+ to Cs^+ (see Table, p. 104). Although the unhydrated lithium ion must be the smallest of these, it moves most slowly because of the great degree of hydration. The relative hydration of the ions may be shown to be in accordance with Fajans' rules, for the readiness of the change $M \rightarrow M^+$ must be the inverse of the readiness with which the reverse reaction $M^+ \rightarrow M$ may be brought about. Hydration, by virtue of the donated electrons from the oxygen, tends to bring about this reverse reaction, which is easiest for the lithium and most difficult for the caesium ions.

While all the compounds of the alkali metals are colourless (unless the acid radical is coloured), they all impart colour to the flame. Flame spectra are due to the excitation of the outermost electrons in the metal atom, which are caused to enter orbitals of higher energy level. When an electron returns to the original orbital the difference in energy is emitted as electromagnetic radiation. In the case of the alkali metals, the energy differences are sufficiently small for the radiation to be in the visible region of the spectrum.

In the majority of their properties, therefore, the alkali metals resemble each other strongly, and differences are mainly of degree. There is, however, rather more difference between lithium and sodium than there is between any other adjacent pair in the group.

s-BLOCK ELEMENTS—GROUP IIA: ALKALINE EARTH METALS

Beryllium	2.2
Magnesium	2.8.2
Calcium	2.8.8.2
Strontium	2.8.18.8.2
Barium	2.8.18.18.8.2
Radium	2.8.18.32.18.8.2

Beryllium is a comparatively rare metal, and partly for this reason its chemistry is not usually studied at intermediate level. It is not an alkaline earth metal in the usual sense, and its chemistry is not closely similar to that of magnesium. Because of its small atomic radius, beryllium is feebly electropositive, so that its compounds are covalent rather than ionic. It compares closely with aluminium, an example of diagonal relationship, and some parallels can also be drawn with the behaviour of zinc.

The remaining metals form a well-graded series, second only to the alkali metals in electropositive nature. The order of increasing electropositive tendency is $Mg \to Ca \to Sr \to Ba \to Ra$, and the same order is observed in comparing most of the properties of the elements. Radium is, of course, far more important for its radioactivity than for its chemical properties, though these do show that it fits into the series.

The metals are light, rather soft and highly reactive. They react with water with increasing vigour, magnesium only on heating, but the others with cold water exothermically. Magnesium alone can be used as a structural metal (usually alloyed with aluminium), as, in the massive form, it suffers surface oxidation only. The metals are obtained by electrolysis of the fused chlorides, not by chemical reduction.

Magnesium shows some differences from the other metals, though most of the differences are of degree. Its hydroxide is much less soluble, and is more easily decomposed by heat. Magnesium gives no flame colour, unlike the other three metals, and it forms organic compounds more readily. (An example of the latter property is provided by the useful Grignard reagents, wherein magnesium shows a strong resemblance to zinc.)

The metals are all exclusively divalent, having in each case two electrons more than the preceding inert gas. The compounds are colourless unless the acid radical is coloured, and strongly ionized, though not as strongly as those of the alkali metals. The compounds are, on the average, more highly hydrated than the corresponding compounds of the alkali metals, though again there is a marked decrease in the degree of hydration as the atomic number increases. This is shown, for instance, by comparing the chlorides: magnesium and calcium chlorides will crystallize with $6H_2O$, but are deliquescent, while barium chloride gives stable crystals containing only $2H_2O$ (cf. also the sulphates).

The oxides and hydroxides of the metals are strongly basic, with no acid properties. The salts are therefore never much hydrolysed in solution. The general order of solubility of the salts is $Mg > Ca > Sr > Ba$. There is a bigger difference between magnesium and the others when comparing the sulphates than with most other salts. The order is reversed with the fluorides (which are sparingly soluble, alone of the halides) as well as the hydroxides mentioned above.

The metals form very few complex salts. Magnesium, which forms the most, again differs somewhat in degree, though the order of gradation is still observed. In almost all cases the complexes are with organic compounds containing oxygen acting as donor, as in the case of the Grignard reagents which coordinate with ether, and then react with other oxygen compounds.

The decreasing tendency to form complexes, as also the decreasing tendency for compounds to be hydrated, is in line with the increasing size of the metallic ions in the order given; this in turn is parallel with the increasing atomic volume (see p. 106). In each period, these elements are second only to the alkali metals in this property, which also explains why they are second, as a group, in the electropositive series.

p-BLOCK ELEMENTS—GROUP IIIB: BORON GROUP

Boron   2.(2,1,)
Aluminium 2.8.(2,1)

Gallium   2.8.18(2,1)
Indium   2.8.18.18.(2,1)
Thallium 2.8.18.32.18.(2,1)

These elements start the p block—in each case, the outer shells of the atoms hold two s electrons and one p electron. All three electrons are available for valency purposes, but are not always used; however, valency one (using only the p electron) is not important until thallium is reached.

Boron is the only non-metal and, as is often the case with the first member of a group, shows more diagonal relationship (with silicon) than group relationship (with aluminium). Boron is a comparatively rare element, very difficult to obtain in a pure state, and very unreactive unless finely divided. It has strong affinity for oxygen—it occurs in the form of borates. It forms stable covalent links with nitrogen, sulphur and the halogens. A number of metals form borides, and the more electropositive metals form borohydrides, containing the BH_4^- ion. Boron forms a considerable number of hydrides; the simplest is B_2H_6, and these compounds were investigated very fully because they do not fit in with simple ideas of valency. They are known as electron-deficient compounds, and are the best known examples, although some other elements also form similar electron-deficient compounds.

GROUP IIIB

The simple tricovalent compounds of boron have only six shared electrons; they react with water or other donor molecules by accepting two further electrons. This explains, for instance, the role of BF_3 as an important catalyst. A covalency of three permits the formation of a giant lattice, and many borates have complicated structures, some laminar, some glass-like and highly refractory. Aluminium, gallium, indium and thallium show a reasonable gradation of properties as rather weakly electropositive metals which, however, become slightly more electropositive with increasing atomic number. This is because the atoms do not increase very much in size compared with the alkali metals or alkaline-earth metals.

Aluminium, by contrast with the other members of the group, is very abundant—the third most abundant element in the earth's crust. Clay, for instance, consists largely of aluminium silicate. This abundance is very unusual for an odd-numbered element (No. 13); all the other abundant elements on earth are even-numbered elements, so there must be some special stability factor in the case of this atom (see chapter on radioactivity and nuclear stability). The metal readily forms an oxide film, and this sometimes resists further attack. Aluminium can be used for structural purposes, although it is attacked by acids (not nitric) and also by caustic alkali. The oxide and hydroxide are amphoteric, but not soluble in water. Compounds are usually covalent; only the fluoride, of the halides, is ionic. As with boron, the compounds readily form complexes with donor molecules, and this accounts for the catalytic activity of aluminium trichloride in organic chemistry. Similarly, the crystalline compounds are frequently hydrated, such as the sulphate and the alums. These compounds are appreciably hydrolysed in solution, because of the weakly-basic nature of the hydroxide.

Gallium, indium and thallium are all low melting-point metals, but have high boiling-points. Like aluminium they tend to form covalent compounds at the group valency, but valency 1 is important in the case of thallium. Gallium shows resemblance to boron in forming electron-deficient hydrides. The heats of oxidation are small, much smaller than in the case of aluminium, and the oxides are therefore much more easily reduced. Gallium hydroxide is amphoteric, but more acidic than aluminium hydroxide—it is readily soluble in ammonia. Indium hydroxide is a weak base with only feeble acidic properties. Thallic oxide (there is no hydroxide) is definitely basic. The metals react readily with the halogens, and the halides again show the increasing electropositive nature of the elements. In the case of gallium, only the fluoride is ionic (cf. aluminium), while with indium the fluoride and chloride are ionic and the bromide is partly ionized. Thallic fluoride is ionic, but the other thallic halides decompose and cannot be studied easily.

The tendency to complex formation increases from gallium to thallium, contrary to what one would expect from the nature of the bases.

The stable valency of gallium and indium is the group valency, 3. Thallium also shows this valency, but is more stable in the monovalent state, in which it resembles the alkali metals in most properties. Monovalency is explained by the 'inert pair' behaviour of two outer shell electrons, a general occurrence in heavy elements (first noted with mercury, immediately before thallium). Indium shows monovalency to some extent, but is oxidized even by water when in this state, whereas thallium is oxidized only with difficulty from valency 1 to 3. Gallium shows the apparent valency 2 in a few compounds, such as the dichloride, which provided a difficulty until the structure was shown to be $Ga^+(GaCl_4)^-$. This means that gallium also shows some 'inert pair' behaviour, although not to the same extent as the other two. This type of behaviour occurs among the heavier elements, and in earlier periods as the group number increases.

It will be seen that there is a gradation of properties from the non-metal boron to the fairly strongly electropositive thallium, although it is not easy to draw general conclusions, owing to the individual behaviour of thallium.

p-BLOCK ELEMENTS—GROUP IVB: CARBON GROUP

Carbon	2.(2,2)
Silicon	2.8.(2,2)
Germanium	2.8.18.(2,2)
Tin	2.8.18.18.(2,2)
Lead	2.8.18.32.18.(2,2)

These elements show a gradation from non-metal to metal, although lead is not highly electropositive. All the elements can complete the octet by simple covalency at the group valency, and the configuration is effectively that of the nearest noble gas.

The elements display allotropy, and this complicates any comparison of physical properties. Carbon for instance, has a 'metallic' form, graphite, which shows electrical conductivity while diamond is an insulator. Silicon and germanium have semiconductor properties. Tin and lead have the conductivities of true metals, although tin has also a non-metallic allotrope stable at low temperature.

Germanium, tin and lead show a valency of two which is increasingly stable in this order. The lower valency is due to 'inert pair' behaviour, which is also found with other *p* block elements. It is this which allows tin and lead in particular to behave as typical metals in forming electrovalent salts; Fajans' rules explain why tetravalent cations are practically unknown, whereas doubly-charged cations are very common.

The elements all form dioxides, and the first two are

(weakly) acidic; the rest are amphoteric, but decreasingly acidic. Lead dioxide is a strong oxidizing agent, because of the stability of valency 2 for lead.

The elements are not very reactive. Carbon is attacked by strong oxidizing agents, silicon by alkalies. Germanium is attacked by oxidizing acids, while tin dissolves in dilute mineral acids and also in boiling alkali. The reactivity of lead is often masked by the insolubility of many of its compounds. The five elements resemble each other in forming volatile hydrides and volatile alkyl derivatives.

Carbon chemistry is governed largely by the great stability of carbon–carbon bonds, whether in chain or ring form. This extends even to multiple bonds between carbon atoms, and multiple bonds are also formed between carbon and oxygen, nitrogen and a few other elements. It is only in the case of a few such compounds that carbon shows a valency lower than 4. Silicon, because of the much weaker Si–Si bond, has a less extensive chemistry, there is no multiple-bond formation and no stable divalent compound. The Si—O—Si linkage is, however, stable, permitting the complex giant molecule structure of some silicates. There is a growing chemistry of silicones and also of other compounds containing both silicon and carbon.

There is a further reason for the stability of carbon compounds, in that carbon has a covalency maximum of four, so that the normal 4-covalent compounds cannot act as acceptors. With silicon the covalency maximum is 6, hence water, a powerful donor, is able to attack compounds such as the tetrahalides, which are completely hydrolysed. Germanium tetrahalides are nearly fully hydrolysed, but with tin and lead an equilibrium is set up; this is because of the greater basicity of these two elements. (The only tetrahalide of lead is the tetrachloride, a sign of the greater instability of lead in the tetravalent state.)

Germanium, tin and lead form monoxides, dihalides and other divalent compounds. In the case of germanium, the compounds are unstable and are reducing agents. The dihalides, though not salt-like, are not very volatile, and appear to be associated by coordination from halogen to metal. As this completes the octet for the germanium atom, it cannot be said that here we have an example of 'inert pair' behaviour.

Stannous compounds are more stable than germanous compounds; they are salt-like, though, again, they are reducing agents. Plumbous compounds are more stable still, and are not reducing agents. In both these cases there is clearly an inert pair of electrons present in the ion, and the plumbous ion is rather more stable than the stannous ion.

It is usually found that the number of complex compounds formed increases with atomic weight in p block groups; in this case this is true as far as tin. For instance, silicon forms complex halides, the salts of fluosilicic acid, H_2SiF_6. Germanium forms similar compounds and also others with chlorine replacing fluorine. With tin, stable complexes are formed with all four halogens. Lead only gives complex halides with fluorine and chlorine, and these are rather unstable, owing to the greater tendency of lead to return to the divalent ionic state.

p-BLOCK ELEMENTS—GROUP VB: NITROGEN GROUP

Nitrogen	2.(2,3)
Phosphorus	2.8.(2,3)
Arsenic	2.8.18(2,3)
Antimony	2.8.18.18.(2,3)
Bismuth	2.8.18.32.18(2,3)

These elements form a graded series, from the totally non-metallic, gaseous nitrogen to the definite metal bismuth, though this element is, of course, not highly electropositive. The biggest change of properties is between nitrogen and phosphorus. Nitrogen is able to form multiple links, including those between two nitrogen atoms, $N{\equiv}N$, and this simple covalent molecule of light weight is very volatile. Phosphorus is apparently much less able to form multiple bonds, and the molecule is tetrahedral P_4, with single bonds between each pair of atoms. This heavier molecule is much less volatile than that of nitrogen; and arsenic, which forms a similar As_4 molecule, is slightly less volatile still. Phosphorus and arsenic have, however, much lower boiling points than those of Group IV elements. Antimony and bismuth are able to form metallic linkages and have boiling points more comparable with those of tin and lead.

Phosphorus resembles arsenic and antimony in showing allotropy; in each case there is a yellow, non-metallic form and a more metallic form. The metallic form is already the more stable in the case of arsenic, while with bismuth there is only the metallic form.

Nitrogen has the curious active nitrogen form, but this can hardly be said to be an allotropic modification.

Nitrogen is not very reactive in the ordinary gaseous form, needing a very high temperature before there is any appreciable reaction with oxygen, although several stable oxides may be made in other ways. Nitrogen will, however, react with metals such as magnesium at moderate temperatures.

Phosphorus is much more reactive to oxidation, and will combine with metals. The remaining elements are steadily less easily oxidized as they increase in atomic weight. This is shown by the temperature to which they must be heated before they burn; phosphorus combines with oxygen at room temperature (in the case of the yellow form) but arsenic, antimony and bismuth must be heated to progressively higher temperatures. Arsenic and antimony resemble nitrogen and phosphorus in forming binary compounds with metals; these compounds are more stable than nitrides or phosphides and are commonly naturally occurring.

Each of the elements has two s and three p electrons in the outer shell, and is three electrons short of the rare gas structure. It is, of course, impossible for stable anions to exist in solution with a triple negative charge, and the easiest method to complete the octet is to form three covalencies. Each of the elements forms compounds of this type, leaving an unshared electron pair. 3-covalent nitrogen is a strong electron donor, and thereby it achieves a completely shared octet. It is impossible for nitrogen to be 5-covalent, as its coordination number is limited to 4, but it can readily lose one electron and become 4-covalent, 1-electrovalent. Phosphorus and the other elements can expand the octet, unlike nitrogen, so becoming 5-covalent. This form, stable with phosphorus, is decreasingly stable in the other elements. With increasing size, however, the elements are increasingly electropositive, and the metallic M^{3+} ions, involving an inert pair, are found to a slight extent with arsenic, appreciably with antimony, and are common in bismuth salts. Finally, just as a large atomic volume favours loss of electrons, so a small atomic volume favours formation of anions, and there is some evidence, in the case of nitrogen only, of the formation of N^{3-}. This may occur in the nitrides of zirconium and titanium and a few other metals, which have an ionic lattice and an electrical conductivity in the fused state. The more likely explanation is that these nitrides are interstitial, and that the conductivity is due to the metal.

All the elements form volatile hydrides of the simple type XH_3. These compounds form a series showing decreasing thermal stability and decreasing basic character. Thus, NH_3 readily coordinates with protons to give the stable cation NH_4^+, because of the strong donor properties of nitrogen. Phosphine shows a much weaker coordination with protons to give the very unstable phosphonium radical PH_4^+. With arsine this property is almost entirely absent. Bismuthine, BiH_3, is very difficult to prepare, and its formation was only proved by tracer technique using a radioactive bismuth isotope. It is the most unstable of the five hydrides. The hydrides are paralleled by a series of alkyl and aryl derivatives; in the case of nitrogen, phosphorus and arsenic, one, two or all three of the hydrogen atoms can be substituted, but in the case of antimony and bismuth only derivatives with all three hydrogens substituted are known. These derivatives are decreasingly stable as we ascend the series.

Nitrogen forms, in addition to ammonia, a number of other hydrides in which there are often at least two nitrogen atoms joined together. This is a sign of the stability of the N—N bond, as well as that of the N—H bond. As the P—P bond is not so stable, we find little evidence of this behaviour, but there is an unstable diphosphine, P_2H_4, an analogue of hydrazine. The remaining elements do not form this type of hydride, but do give alkyl- and aryl-substituted compounds with two linked arsenic, antimony or bismuth atoms.

Reference to the many oxides of nitrogen has already been made. These are unusual in that the anhydrides of the common nitrogen acids, N_2O_3 and N_2O_5, are not the most stable oxides, and the stable oxides are neutral rather than acidic. Nitric acid differs from orthophosphoric acid not only in formula, but also in being less stable to heat, a much stronger acid and a strong oxidizing agent. Furthermore, all its salts are soluble, whereas most phosphates are insoluble. Nitrous acid is a weak acid, and differs from phosphorous acid in being much less stable.

Phosphorus, arsenic and antimony oxides and oxyacids form a well-marked series. Each element forms a trioxide and a pentoxide; the former exists in the dimeric form. Phosphorus trioxide reacts vigorously with water, and As_2O_3, Sb_2O_3 dissolve in water to give weak acids of decreasing strength in the order given above. Phosphorus trioxide has no basic properties, but arsenious and antimonous oxides show solubility in acids, and antimony can form a variety of salts. Phosphorous acid is a reducing agent, but the corresponding acids of arsenic and antimony have no great tendency to be oxidized. All three acids show the usual property of polyhydroxy acids of forming a series of condensed acids (or their salts) on heating the free acids or their metal derivatives.

Phosphorus readily oxidizes to the pentoxide (the

dimeric form), which reacts vigorously with water to give phosphoric acid; the final product is the orthoacid, H_3PO_4. Arsenic and antimony both need oxidizing agents to convert them to the pentoxides. Arsenic pentoxide is very soluble in water to give H_3AsO_4. This acid is very similar to phosphoric in the solubilities and structures of the corresponding salts, and in the series of condensed acids and their salts obtainable on heating. One or more phosphorus or arsenic acid residues can replace molybdenum or tungsten oxides in their acids to give heteropoly acids.

Antimony pentoxide is slightly soluble in water to give a very weakly acid solution, and colloidal antimonic acid may be obtained by hydrolysis of antimony pentachloride. The product is soluble in mineral acids. The antimonates differ from the arsenates and probably resemble the tellurates more closely, as they have been shown to have the structure $M[Sb(OH)_6]$.

Bismuth trioxide differs sharply from antimony trioxide in being only basic, and it gives no definite compound with water. It shows some signs of further oxidation under stringent conditions, when fused with caustic soda and sodium peroxide. The product, known as sodium bismuthate, is not a pure compound and can hardly be compared with antimonates.

The elements all form trihalides, which show the gradation of the group better than the oxides or oxyacids. Thus, with the exception of nitrogen trifluoride, all the trihalides are hydrolysed to a greater or lesser extent. The other nitrogen trihalides are immediately hydrolysed completely, giving ammonia and hypohalous acid. (There is of course no hypofluorous acid.) The weakness of the nitrogen-halogen link is further shown by the explosive nature of the halides. Phosphorus trihalides are also hydrolysed completely, but to give phosphorous acid and halogen acid. The action is slow with the trifluoride, and rapid with the other trihalides. Arsenic trihalides are hydrolysed in a similar way, but the action is reversed by the addition of halogen acids. With antimony and bismuth the hydrolysis, again reversible, stops at the oxyhalide stage, and the oxyhalides are salt-like in character. The more electropositive nature of bismuth is further shown by its trifluoride, which is electrovalent, and by its trichloride, which is partly ionized.

Nitrogen can form no pentahalides, while phosphorus forms a pentafluoride, pentachloride and pentabromide. Arsenic forms only a pentafluoride, while antimony forms also a pentachloride; bismuth forms no pentahalides. On the whole, therefore, the covalency 5 decreases in stability from phosphorus to bismuth.

Phosphorus pentahalides are rapidly hydrolysed in two stages, via the oxyhalide. Arsenic and antimony pentafluorides are stable, but are fluorinating agents. Antimony pentachloride will form a hydrate; the pentachloride largely dissociates on heating.

Arsenic, antimony and bismuth all form a number of complex compounds of several types. These atoms can act as donors when in the trivalent state, just as nitrogen does. They can also act as acceptors in the same valency state (unlike nitrogen), and here there will be a completely shared octet in addition to the unused pair, which is therefore acting as an inert pair. Thirdly, they can act as acceptors when pentavalent, but here there is no inert pair but a shared group of twelve electrons. An example is afforded by the complex fluoride, $K[SbF_6]$.

p-BLOCK ELEMENTS—GROUP VIB: OXYGEN GROUP

Oxygen	2.(2,4)
Sulphur	2.8(2,4)
Selenium	2.8.18.(2,4)
Tellurium	2.8.18.18.(2,4)
Polonium	2.8.18.32.18.(2,4)

These elements have the same relationship to each other as the corresponding elements in Group V, but are more electronegative. They all show allotropy, and selenium, tellurium and polonium have 'metallic' allotropes, but are not regarded as metals. The chemistry of polonium has not been studied extensively, as the element is exceedingly rare. All the elements will combine with a large number of metals, especially if finely divided, usually on gentle heating; they are second only to the halogens in their reactivity with metals. The binary compounds differ from the majority of halides, in that they are normally insoluble or sparingly soluble. Further, when the compounds are soluble they do not readily ionize strongly, because the divalent anion is not very stable. By Fajans' rules, the smaller the anion the less likely it is to become covalent, because the attracted electrons are closer to the positive nucleus. We might therefore expect oxygen ions; but the position of oxygen in solution is complicated by hydrolysis, and O^{2-} cannot exist in solution. The remaining elements show the expected order of ionization, the sulphides being the most strongly ionized of the binary compounds.

Many solid metallic oxides are ionic, as shown by their crystal structure, high melting point and electrical conductivity in the fused state (when this can be measured).

The hydrolysis of O^{2-} in solution gives of course the much more stable hydroxyl ion, stable because only one electron has to be held electrovalently, by an anion which is still small. The first ionization of H_2S, and the hydrolysis of soluble sulphides, gives the analogous SH^-, and this is paralleled by the other elements, but to a decreasing extent.

The elements combine with hydrogen to give volatile hydrides. Water differs from the other hydrides in a number of ways; it is much less volatile (because of association) than would be expected from the position of oxygen. The compound is much more thermally stable than the other hydrides, which decrease in stability through the series. Hydrogen sulphide is a stronger acid than water, and the acid strength increases through the series. The explanation of the weakness of ionization of water may again be hydrogen bonding, which accounts for association, and is for some reason much stronger between hydrogen and oxygen O—$H \cdots O$ than between hydrogen and sulphur, S—$H \cdots S$. The hydrides of the elements from sulphur throughout the group all have evil odours. Water is a powerful solvent, promoting ionization (although scarcely ionized itself), partly because of its high dielectric constant and partly because it can act as donor (through oxygen) and effectively as acceptor (through hydrogen bonding). Hydrogen sulphide is a very poor solvent, and does not promote ionization; it has a low dielectric constant.

Oxygen has a slight power of forming links with itself, as in hydrogen peroxide and the metallic peroxides. These normally evolve oxygen readily on heating or treatment with acids. The corresponding sulphur compounds are known, and sulphur has a greater power of forming chains, shown by the number of polysulphides known, some fairly stable. Selenium shows little tendency to form polyselenides, though H_2Se_2 may exist, but a few unstable polytellurides are known.

All the elements give dioxides, which show, however, a divergence of properties. Ozone, O_3, cannot be compared with the others, as it is more in the nature of a peroxide. Sulphur dioxide is a gas, very soluble in water, giving the rather weak sulphurous acid, which can be completely decomposed by heat. Sulphur dioxide is both a reducing agent, being oxidized to the trioxide, and an oxidizing agent, being reduced to sulphur. Selenium dioxide is a solid, soluble in water to give selenous acid, weaker than sulphurous. Selenium dioxide is an oxidizing agent, useful in organic chemistry for its selective action. It does not seem to have any reducing power, as the trioxide has no separate existence. Tellurium dioxide is a solid which is almost insoluble in water, but soluble in alkalies to give tellurites; it also forms addition products with strong acids, and solutions in acids contain a low concentration of Te^{4+} ions, so the dioxide is amphoteric. It is also an oxidizing agent, and can be used instead of selenium dioxide. Tellurites in alkaline solution are oxidized to tellurates, but not in acid solution. Tellurous acid is weaker than selenous acid, and is also weaker than carbonic acid.

Sulphur and tellurium give trioxides, but selenium is apparently more reluctant to combine with oxygen, just as bromine, in the same period, is unlike the other halogens in not giving perbromates. The three elements give the corresponding acids, but whereas sulphuric and selenic acids and their salts resemble each other, telluric acid shows a number of differences. The normal free acid is $Te(OH)_6$ and this gives some salts and esters. There is a strong tendency to give condensed acids, and tellurates are not usually isomorphous with sulphates and selenates. The acid readily becomes colloidal, and is more readily reduced than either selenic or sulphuric acid; as an acid it is much weaker than the other two, while selenic acid is slightly stronger than sulphuric acid.

While there are a number of similarities between oxygen and sulphur, which form similar compounds, the difference between the two is brought out when comparing the halogen compounds. Oxygen unites with the halogens to give highly unstable, reactive compounds such as the monoxides and dioxides, with oxygen probably exerting its normal valency of 2. Sulphur, selenium and tellurium form hexafluorides: once again fluorine compels elements to exert their highest valencies. Sulphur and selenium hexafluorides are covalent compounds, stable because in them the elements reach their covalency maximum. Tellurium hexafluoride is hydrolysed by water: the covalency maximum for tellurium is 8. With the other halogens, lower valencies are the rule. Sulphur forms mono- and dihalides, which are volatile, unstable liquids readily hydrolysed. Selenium and tellurium give a number of tetrahalides, evidence for the 'inert pair' of electrons. These, and the lower halides, are hydrolysed, not very readily in the case of tellurium halides, which, although volatile, have a slight conductivity.

The valency states which these elements can show are as follows.

(*1*) Electrovalency of two, covalency of two, completing the octet in each case.

(*2*) Covalency of four, with an inert pair, in the case

GROUP VIB

of selenium and tellurium, but not definitely with sulphur. (This involves a shared octet in addition to the inert pair.)

(*3*) Covalency of four, in the case of sulphur, by coordination as in $S{\displaystyle{\diagup O \atop \diagdown O}}$, involving six shared and two unshared electrons.

(*4*) Covalency of 6, with 12 fully shared electrons in sulphur, selenium and tellurium, and with 8 shared electrons by coordination of sulphur in the form $O=S{\displaystyle{\diagup O \atop \diagdown O}}$.

(*5*) Electrovalency of 4 in the case of tellurium, Te^{4+}, again with an inert pair.

(*6*) In addition, oxygen and sulphur in the 'oxonium' and 'sulphonium' compounds become 3-covalent by losing one electron. The simplest example of this is the hydrated hydrogen ion in all acids, $(H_3O)^+$. There are a number of organic oxygen and sulphur compounds showing this behaviour, with some of the hydrogen atoms substituted by alkyl groups.

These diverse valencies make the chemistry of this group of elements as complicated as that of any transition group (where variation of valency is brought about by electrons being transferred from the penultimate group to the outermost group). In this case, the simple octet is not so readily attainable that it dominates behaviour, and instead the octet readily becomes more fully shared, or expands, as circumstances permit or dictate.

Because of the stability of sulphur chains, sulphur gives a whole series of oxyacids containing more than one sulphur atom. The sulphur can be partly, but usually not wholly, substituted by selenium or tellurium. The oxyacids of other elements, such as carbon, arsenic, antimony, etc., in each case give thio-analogues, in which some of the oxygen is replaced by sulphur.

The various oxyacids of sulphur, selenium and tellurium give organic derivatives, not only the esters, but also compounds in which alkyl or aryl groups replace hydroxyl groups of the acids. In these and other organic compounds, the elements form stable links with carbon atoms.

It will therefore be seen that, while these elements have a number of properties in common, the two which are most closely similar are sulphur and selenium.

p-BLOCK ELEMENTS—GROUP VIIB: THE HALOGENS

Fluorine 2.(2,5)
Chlorine 2.8.(2,5)

Bromine 2.8.18.(2,5)
Iodine   2.8.18.18.(2,5)
Astatine 2.8.18.32.18.(2,5)

These elements have structures which are one electron short of an inert gas structure, and are therefore very reactive and highly electronegative. The elements themselves form covalent, volatile molecules showing an increase in melting and boiling points as the molecular weight increases. They have unpleasant smells and are irritant, corrosive poisons.

As, by Fajans' rules, electrovalency is favoured by small size of anions, fluorine is the most active element known in forming simple anions, and there is a gradual decrease of reactivity through the halogens as the size of the atom increases. This is clearly shown by the oxidation of halide ions by lighter halogens, e.g. $2Br^- + Cl_2 \rightarrow 2Cl^- + Br_2$. The lighter halogen is more electronegative than the heavier.

Fluorine attacks most elements at room temperature or on very slight warming, often violently; the element concerned usually exerts its highest valency with fluorine, and the other halogens are included in this statement (e.g. IF_7), but not always the transition metals. Metal fluorides are often ionic, even when other halides of the metal are covalent (e.g. Al, Hg); in any one series of halides the fluorides are usually ionized to the greatest extent. Fluorides often show solubility differences from other halides; for instance, silver fluoride is the only soluble silver halide, though there are more cases of sparingly soluble fluorides when other halides are soluble (e.g. calcium, magnesium).

Chlorine attacks almost as many elements as fluorine, and the vigour of the reaction is second only to that of fluorine. Reactions of metals with bromine and iodine are usually slightly less vigorous.

When comparing the four halides of any one metal, it is found on average that the degree of hydration increases with the atomic weight of the halogen. This is because the donor properties of the anion increase with its size.

The halogen acids are covalent, volatile compounds which are very soluble in water to give highly ionized solutions, except in the case of hydrofluoric acid. This is less volatile and less ionized, because of its power of association by hydrogen bonding; for the same reason it can give acid salts, alone of the halogen acids. For the explanation of the complete change from covalency to electrovalency of the hydrogen halides on dissolving we must look to the nature of water rather than to that of the halogens. The covalent hydrogen halides may perhaps be compared with the halogen molecules themselves, and so used to justify the consideration of hydrogen as a halogen, which is, however, less electronegative than iodine, although so light in weight. The dipole moments of the gaseous hydrogen halides decrease steadily from the high value found for HF, and show that, although the bond consists of two shared electrons, these are not equally shared by hydrogen and halogen; the latter has the greater share, and is the negative pole. A very similar distribution of electrons occurs in the simple interhalogen compounds, where the lighter halogen is found to be the negative pole.

Fluorine forms no oxyacids, although the element does form four oxides; these are unstable, reactive compounds, and confirm the reluctance of fluorine to combine with oxygen. There are comparatively few oxyfluorides known, although other oxyhalides are fairly common. Chlorine forms a series of oxyacids with from one to four oxygens per chlorine atom. Perchloric acid is stable and a strong acid, showing that the fully shared octet is a stable structure for chlorine. Bromine will give oxyacids with up to three oxygens, but not more, although iodine gives periodic acid like perchloric acid. Chlorine forms several oxides, which resemble fluorine oxides in being unstable, vigorous oxidizing agents. Bromine forms, again with reluctance, three oxides which are unstable. The oxides of iodine are rather more stable than those of the other halogens. This indicates the slightly 'metallic' tendency of iodine, which is further shown by the fact that some of its compounds give evidence of the presence of positively charged iodine, e.g. ICl.

The halogens form a large number of organic compounds; but the fluorides differ in properties from the other halides in some respects. The highly fluorinated organic compounds are very stable towards hydrolysis and other replacement reactions; fluorides do not always react with magnesium in the way that other organic halides do; and the highly fluorinated compounds are less toxic than the corresponding chlorides.

These and other properties show that there is more difference between fluorine and chlorine than between the other neighbouring halogens.

Astatine has been studied in sufficient detail to show it has the properties one would expect. It behaves like iodine in most respects, but with more tendency to assume the positive form. It is intensely radioactive; even the most stable isotope has a half-life period of only a few hours.

p-BLOCK ELEMENTS—GROUP 0: THE NOBLE GASES

Helium 2
Neon 2.8
Argon 2.8.8
Krypton 2.8.18.8

Xenon 2.8.18.18.8
Radon 2.8.18.32.18.8

The 'rare' or 'inert' gases were separated as a result of Lord Rayleigh's work just before the turn of the 19th century; the name 'rare' is particularly inappropriate for argon, because it forms nearly 1 per cent of the atmosphere. Although the Periodic Table had been constructed before the existence of these gases was realized, no great modifications were necessary, as the gases fit in between the extremely electronegative halogens and the extremely electropositive alkali metals, so providing a 'buffer state'.

For many years, these elements could be said to have no chemistry; until recently, no definite stable compounds of conventional type had been confirmed. However, in 1963, fluorides of xenon (XeF_4) and krypton (KrF_4) were obtained as crystalline solids, and also a complex fluoride, $Xe^+(PtF_6)^-$. It is in accordance with general theory that the larger atoms should react, if any do. The (first) ionization potential of xenon, in particular, is not as high as that of many metals.

Later work has shown that xenon forms other

fluorides and many complex fluorides; it also forms unstable oxides XeO_3 and XeO_4.

Previously, hydrates of argon, krypton and xenon have indeed been claimed, and dissociation pressure figures quoted; these are all over 1 atmosphere at 0°C, so that these hydrates would only be stable at low temperatures. It is, again, reasonable to suppose that the donor powers would increase with atomic number, and xenon hydrate is in fact stated to be the most stable. Argon and krypton do give clathrates, such as that formed with hydroquinone, but these should not be compared with compounds formed with normal valency links. The inert gas atoms are locked or imprisoned in the interstices of the hydroquinone crystals, and cannot escape, although under considerable pressure, until the compound is melted or dissolved. The lighter gases have atoms too small to be held, and the heavier ones have atoms too big to be included in hydroquinone lattices.

The noble gases are monatomic under all conditions, and a determination of the vapour density has been the way to measure the atomic weight (except for the physical method in the mass spectrometer). They are valuable because of their inertness in a number of applications, such as for filling lamp bulbs, and for allowing welding of readily oxidizable metals by providing an inert atmosphere.

A consideration of the place of these gases in the Periodic Table was of very considerable help in building up the modern theories of atomic structure and electronic interpretation of valency. It was obvious that the structures of the inert gas atoms must be of great stability, and the valency of an element (in the short periods at least) was equal to the number of places before or after an inert gas. In these cases it could be seen that reactions took place to give atoms an effective structure of greater stability, similar to that of an inert gas. Each inert gas has an electron structure which is extremely stable; except in the case of helium the outer shell contains eight electrons. Helium has two electrons only, in the first shell, and this is the maximum possible for that shell.

Radon is important for its position in radioactive series, and also for its own radioactivity, which is employed in radiotherapy. It was formerly known as 'emanation', and different isotopes are associated with radium, thorium and actinium. Helium is also associated with radioactive minerals, because α particles ejected during disintegration are charged helium atoms. This suggests that the arrangement of two neutrons and two protons is a stable unit which may figure in the nucleus of heavy atoms.

For these various reasons, therefore, the discovery of the family of elements almost devoid of chemical properties has been most helpful in promoting and stimulating important advances in chemical theory. They have been well named the 'noble gases'—a name which has now been revived—since they can no longer be considered completely inert.

d-BLOCK ELEMENTS—GROUP IIIA: SCANDIUM GROUP

Scandium 2.8.(8,1)2
Yttrium 2.8.18.(8,1) 2
Lanthanum 2.8.18.18.(8,1) 2
Actinium 2.8.18.32.18.(8,1) 2

These elements form the first group of the *d* block, and all have one *d* electron in their penultimate shell; the outer shell contains two *s* electrons in each case. The single *d* electron does not need much energy to be promoted to the outer shell, whereby it is available for valency purposes, and the valency is three throughout the group.

These elements, all rare, are metals which are more electropositive than aluminium and become slightly more electropositive on descending the group. Scandium oxide and hydroxide are weakly basic, and lanthanum oxide is almost as basic as calcium oxide. Actinium oxide is similar to lanthanum oxide, but slightly more basic still. There is a correspondng decrease in the degree of hydrolysis of compounds and in the readiness to form complex compounds in line with the slowly increasing atomic radius. The differences between lanthanum and actinium chemistry are slight, and it is usual to employ ion exchange processes to bring about effective separations.

Lanthanum is followed by the first of the *f* block series (lanthanons), in which the fourth shell builds up from 18 to 32, and actinium is followed by the second series (actinons), in which the fifth shell similarly

GROUP IIIA

builds up from 18 to 32. These are known as inner transition elements, whereas the d block elements are known simply as transition elements. However, the scandium group do not show typical transition properties fully developed; variable valency, coloured ions and high catalytic activity are all absent. There is some tendency to form complex ions, as mentioned, and the spectra are characteristic of transition metals.

d-BLOCK ELEMENTS—GROUP IVA: TITANIUM GROUP

Titanium   2.8.(8,2) 2
Zirconium 2.8.18(8,2) 2
Hafnium   2.8.18.32.(8,2) 2

In the older forms of the Periodic Table, this subgroup includes thorium, which is now classed as the second member of the actinons. There is no more justification for including thorium than for including cerium (which has also appeared in Group IV in some Tables) or for showing transition elements in several different groups in accordance with their several different valencies. Omitting thorium, however, introduces the difficulty of showing group tendencies from the behaviour of effectively only two elements, as hafnium is so similar to zirconium in its properties.

The electron structure of titanium is 2.8.(8+2)2. It shows a very unstable valency of two, and is a strong reducing agent in valency three, so that the most stable state is when the four possible valency electrons are used. The structures of zirconium and hafnium are similar, and their lower valency states are even less stable than in the case of titanium.

The metals have very high melting points, and at high temperatures combine vigorously with oxygen, nitrogen and carbon. They are therefore very difficult to prepare in a pure state, the best method being the pyrolysis of the (volatile) tetraiodides on a tungsten filament. Because of the affinity for oxygen and nitrogen at high temperatures, titanium is a valuable additive to steels, when quite small quantities have a big effect on strength. The action is the removal of gaseous impurities, known as 'scavenging'; zirconium behaves similarly, and the two metals are also used in high-vacuum equipment for removing the last traces of air.

The metals are very inert at ordinary temperatures; they are not attacked by concentrated acids in the cold, but are attacked by a limited number of acids when heated. Moderate heat is also necessary before the halogens will react.

Hafnium, No. 72, coming as it does immediately after the lanthanons, shows markedly the effect of the lanthanide contraction (p. 81) and has the same atomic and ionic radius as zirconium, within experimental error. This accounts for the two elements being extraordinarily similar in properties, as already noted; they are more closely related than two neighbouring lanthanons, and consequently are very difficult to separate. The method now favoured is ion exchange, using the tetrachlorides in methanol.

The three elements form carbides such as TiC, which are very hard, very refractory and very resistant to chemical attack. These compounds resemble silicon carbide, but are not paralleled in Group IVB.

The three metals form nitrides by strongly heating in nitrogen, and these are stable and infusible. They have the sodium chloride lattice and show some electrical conductivity, so that they may contain the nitride ion, N^{3-}. However, the facts may also be interpreted to mean that the compounds are interstitial.

The elements form dioxides which are very similar to each other; they are stable, white, refractory oxides which are very resistant to attack by acids, though they are slowly attacked by fused alkalies. The corresponding hydroxides form a series which is decreasingly acidic. They are often colloidal substances which readily lose water. Titanium hydroxide, for instance, reacts with fused alkalies to give titanates, similar to silicates but more hydrolysed in solution. Titanium hydroxide, when freshly precipitated, also dissolves in acids to give titanic salts. Zirconium and hafnium hydroxides are more basic than acidic, as they dissolve readily in acids but only with difficulty in alkalies.

The tetrahalides are all covalent, volatile substances which are much hydrolysed even in acid solution. Zirconium and hafnium show slightly more readiness to form salts of oxyacids than does titanium.

Titanic compounds can be reduced fairly easily to valency state 3; the resulting compounds are strong reducing agents and are only stable under hydrogen. Divalent halogen compounds have also been prepared, also di- and trivalent halides of zirconium and hafnium. These are all unstable, vigorous reducing agents reacting immediately with air or water. The compounds are coloured, as is usual with transition metals.

The metals all form a number of complexes, chiefly with oxygen-containing ligands, particularly the chelated β-diketone derivatives. The number of complexes tends to fall off from titanium to hafnium.

On the whole, therefore, the properties support the view that there is some increase of electropositive nature from titanium to hafnium. This second group of the d block displays most of the typical transition metal properties.

d-BLOCK ELEMENTS—GROUP VA: VANADIUM GROUP

Vanadium 2.8.(8,3) 2
Niobium 2.8.18.(8,3) 2
Tantalum 2.8.18.32.(8,3) 2

Although traditionally this subgroup includes protoactinium, that element will not be included here, for the reasons discussed in Group IV. It is the third member of the actinons.

Vanadium is comparatively scarce, and the other elements are rarer still. The metals are difficult to extract, and are normally available as master ferro-alloys which are used to prepare a series of valuable alloy steels. Small quantities of vanadium act, like manganese or titanium, as oxygen scavengers, while larger quantities confer toughness. Niobium and tantalum have great resistance to chemical attack, a property shared by their alloys, which are also valuable in applications where resistance to high temperature is necessary.

Vanadium is attacked by oxidizing acids and by halogens; it reacts with oxygen and sulphur at high temperatures. Niobium is attacked by halogens but not appreciably by acids, and tantalum is still more inert—it reacts only on considerable heating.

The metals form a typical transition series, showing variable valency, coloured compounds (particularly in the lower valency states) and catalytic activity, notable with vanadium pentoxide. Vanadium can show any valency from 2 to 5. In valency states 2 or 3 it forms basic oxides which give rise to a number of coloured salts, which are reducing agents. In valency state 4 it is amphoteric, giving vanadyl salts with the radical VO^{2+} and also unstable vanadites. Valency 5 is the most stable; vanadium pentoxide is acidic, giving vanadates (usually colourless), but it is also weakly basic, and absorbs strong acids to give vanadyl salts containing the VO^{3+} vanadyl radical.

Niobium and tantalum are quite similar to each other, always occur together and are difficult to separate. The reason for their similarity is the effect of the lanthanide contraction which makes their atomic volumes very nearly the same. Niobium forms an oxide NbO, and it can exist also in the trivalent form, when it shows strong reducing powers. It is not certain whether it can show valency 4, and its most common valency is 5. The pentoxide is amphoteric, but much more basic than acidic, and resistant to attack by any agent except at high temperature. Tantalum is similar in most details; the pentoxide is even less easily attacked than niobium pentoxide, but the freshly precipitated hydrated oxide is soluble in both acids and fused alkalies.

Considering the three metals, which are all stable in the pentavalent form, it will be seen that the lower valencies become less stable with increasing atomic number. The few compounds of lower valency are not ionic in the case of niobium and tantalum, and these metals have 'noble' properties as have the remaining transition elements of the second and third long periods.

The vanadates, niobates and tantalates formed from the pentoxides and alkali have some properties in common with the phosphates and arsenates, but there are also some differences. There are in all cases condensed or isopoly acids, but these are more readily interconverted in the case of vanadium, niobium and tantalum oxyacids, which therefore show some resemblances to the chromium oxyacids. The simpler acid radicals exist in strongly alkaline solution, and as the pH decreases the acids give the more condensed forms. In strongly acid solution the oxides may be precipitated, and in the case of vanadium, the oxide may dissolve in excess acid.

Vanadium forms a pentafluoride, niobium and tantalum form all the pentahalides. All these compounds are volatile, and are hydrolysed by water, giving oxyhalides in the first place with vanadium and niobium. Tantalum does not form oxyhalides as easily as the other two, and this is the basis of the usual method of separation from niobium. Vanadium forms an oxychloride and oxybromide of the type VOX_3, in addition to an oxyfluoride. These are volatile and presumably covalent, though some of the vanadyl compounds are salt-like, with an appreciable conductivity in solution.

The number of complexes given by these metals in general falls off with increasing atomic number. This, and the decreasing acidity of the oxides, would indicate increasing electropositive tendency on descending the group; but it is impossible to regard niobium and tantalum as more electropositive than vanadium, in view of their unreactivity and absence of simple ionic compounds. This conflicting evidence is what would be expected when the Periodic Table is considered as a whole. (See the section so entitled, and the diagram given there, p. 86.)

Group IIIA elements have the electronic structures which make them the first transition group, but they do not show many chemical transition properties. Group IVA elements show most of the transition properties, and these are fully developed by the time

THE PERIODIC TABLE

that Group VA is reached. Besides the usual transition properties of variable valency, catalytic activity, diamagnetic properties and coloured compounds, the further properties of close similarity, nobility, stability of higher valencies and absence of ionization in lower valencies are developed in the second and third elements of each group. These make possible a close comparison of 'horizontal' neighbours.

d-BLOCK ELEMENTS—GROUP VIA: CHROMIUM GROUP

Chromium   2.8.(8,5) 1
Molybdenum 2.8.18.(8,5) 1
Tungsten   2.8.18.32.(8,4) 2

In the chromium and molybdenum atoms, the d shell is half-filled with 5 electrons, and there is only a single s electron. In the tungsten atom, energy considerations dictate that two s electrons are retained, and there are only 4 d electrons. This fine difference of atomic structure does not affect the chemistry of the elements significantly.

Chromium, molybdenum and tungsten are transition metals with close resemblances to their neighbours in Subgroups VA and VIIA; the pattern due to the lanthanide contraction is continued, so that molybdenum is more like tungsten than chromium. The melting points, and also resistance to chemical attack, increase with atomic weight; the metals have considerable 'nobility'. They are valuable constituents of alloy steels, to which they are added as ferroalloys. The alloys acquire some of the corrosion resistance of the pure metals.

Chromium shows valency states 2, 3 and 6, valency 3 being the most stable. Chromous compounds are strong reducing agents, being oxidized even by air, unless stabilized by complex formation. In valency state 3, chromium resembles aluminium; the sesquioxide is amphoteric, although more basic than acidic. Chromic compounds ionize to some extent, but are often complex, and often show isomerism with varying degrees of ionization. In valency state 6, chromium becomes acidic and strongly oxidizing. The trioxide is volatile, covalent and soluble in water to give a strong acid which is much used in organic chemistry as an oxidizing agent. The salts, however, when in alkaline solution are not as strongly oxidizing as permanganates. The acid and salts show a strong tendency to condense, forming isopoly acids as the pH of the solution drops. Many oxyacids of elements in other groups show a similar tendency to give condensed forms, though this often requires heat and the equilibrium is seldom as easily attained as in Group VIA.

Molybdenum shows valencies from 2 to 6. There are a few compounds of valency 2, which have reducing action, and there is a general 'preference' for higher valencies. Strong reducing agents usually give valency 3, and at this valency the oxide and hydroxide, insoluble in water, are basic. Representing valency 4, there is a disulphide (the chief ore of molybdenum) and a dioxide, an inert, refractory compound. The tetrahalides are known, and there are also a number of complex compounds. In valency state 5, there is a pentachloride, and there are a number of molybdenyl compounds. Of the compounds having the group valency, the trioxide is a white solid almost insoluble in water, but with acid properties. The molybdates, like the chromates, readily condense as the pH drops. In addition to the isopoly acids, heteropoly acids can also be formed by the inclusion of a residue of another acid such as phosphoric, silicic or arsenic acid. In contrast to chromic acid, molybdic acid is not an oxidizing agent, and it shows weak basic properties by reacting with strong acids to give molybdenyl compounds.

GROUP VIA

Tungsten is quite similar to molybdenum in most respects. There are even fewer compounds at valencies 2 and 3; at valency 4 there is a disulphide and dioxide similar to those of molybdenum, and also tetrahalides. There is both a pentachloride and a pentabromide, and again tungstyl compounds are formed. At valency 6, the trioxide and tungstic acids, both isopoly and heteropoly, are similar to the corresponding molybdenum compounds. Tungsten forms three hexahalides, while molybdenum forms only the hexafluoride; these compounds are volatile.

The relationships in Subgroup VIA are very similar to those in VA. There is the same decreasing importance of lower valencies with increasing stability of the group valency, and a similarity between second and third metals, rather than first and second. On the other hand, there is a greater tendency for partial hydrolysis to occur, giving molybdenyl and tungstyl compounds, at the higher valencies, whereas tantalum compounds resist partial hydrolysis.

d-BLOCK ELEMENTS—GROUP VIIA: MANGANESE GROUP

Manganese 2.8.(8,5) 2
Technetium 2.8.18.(8,6) 1
Rhenium   2.8.18.32.(8,5) 2

While manganese is quite abundant (about tenth in order of abundance), technetium has no stable isotope and is unknown in nature, and rhenium is very scarce, with no good ores. It has, however, been studied in some detail, and enough is known of the chemistry of technetium (obtained by alpha bombardment of molybdenum) to know that it resembles rhenium in most respects.

Manganese is an important constituent of alloys, both ferrous and non-ferrous. Small quantities are added to steel as a 'scavenger', because of its affinity for oxygen; larger quantities improve the toughness and strength of steel.

Manganese is a reactive metal, rather similar to magnesium in this respect. It will decompose water, quite rapidly on heating, and liberate hydrogen from dilute acids, including nitric acid, giving divalent manganous salts. It will burn if heated in nitrogen, to give the nitride, again like magnesium.

In other respects, however, manganese differs completely from magnesium: it is a typical transition metal, showing valencies from 1 to 7. Only a few complex monovalent compounds have been isolated, and the most stable valencies are 2, 4 and 7. In valency state 2, it is basic and forms typical salts.

GROUP VIIA

The electronic structure of the manganous ion is $2.8.(2.6.5)^{2+}$ so that the third shell is half way from 8 to 18, a fact which apparently accounts for the stability, and manganous salts are reluctant to form complexes.

Trivalent manganese is still basic, and forms a few salts, though these are unstable, tending to give manganous salts on heating. They can be stabilized by complex formation. Manganese dioxide is amphoteric and an oxidizing agent, and gives rise to the manganites and a limited number of complex compounds. Hexavalent manganese has no basic properties; the manganates are strong oxidizing agents, but are very ready to disproportionate to Mn^{vii} and Mn^{iv}, except in alkaline solution. In valency state 7, manganese forms the volatile, explosive heptoxide, a very strong oxidizing agent. This gives rise to permanganic acid (known in solution) and the permanganates, reasonably stable salts which are strong oxidizing agents in acid or alkaline solution.

Metallic rhenium is less reactive than manganese. It is susceptible to oxidation with oxygen, halogens, or oxidizing acids, but does not readily react with other acids. Rhenium shows every valency from 1 to 7, but compared with manganese, the lower valencies are less stable. There are very few compounds showing valencies 1 or 2, but valency 3 is a little more stable than in the case of manganese. Although the sesquioxide is basic, the compounds of trivalent rhenium do not appear to ionize, and are often complex. Otherwise, oxidation to a higher valency state becomes easy, and there is for instance no trifluoride. Rhenium dioxide resembles manganese dioxide in being amphoteric, but is not an oxidizing agent. It will combine with further oxygen, for instance on warming in air, to give the heptoxide. Tetravalent rhenium gives rhenites, a tetrafluoride, and a series of complex halides of the type $M_2(ReHal_6)$ with all four halogens. Manganese forms far fewer complex halides of this type, showing that rhenium is more stable than manganese in the tetravalent state.

Most pentavalent rhenium compounds disproportionate readily, giving Re^{vii} and Re^{iv}; the hexavalent compounds show a still greater instability and disproportionate in the same way. They are even less stable than hexavalent manganese compounds, possibly because of the greater stability of heptavalent rhenium. This state is represented by the heptoxide and the perrhenates. The oxide is volatile and can be distilled unchanged, in contrast to manganese heptoxide, which readily explodes; rhenium heptoxide is not an oxidizing agent—its reduction involves heating with hydrogen, carbon monoxide, or a similar reducing agent. It is white or yellow in colour, and perrhenates are similarly pale-coloured, in contrast to the permanganates. The perrhenates resemble permanganates and perchlorates in the solubilities of corresponding salts: with the exception of the sodium salts, even the alkali metal salts are sparingly soluble. Perrhenates, like the heptoxide, are much more stable to heat than permanganates (or perchlorates). Potassium perrhenate may even be distilled unchanged at over 1300°C. This property of stability at the highest valency is employed in the separation of rhenium from natural sources, where it rarely occurs to a greater extent than a few parts per million.

While rhenium shows a number of similarities to manganese, the main difference is this greater stability of valency 7 (and, to a smaller extent, valency 4). This is in line with the change in properties of each of the later A subgroups, from IVA onwards through the Periodic Table. Of the usual transition properties, variable valency has been dealt with at some length; coloured compounds are found at most valency states (though manganous compounds are very pale, and perrhenates have been mentioned as an exception). Catalytic activity is notable with manganese dioxide and several other manganese compounds, and is also found with finely divided rhenium. It is seen therefore that these metals have many points of similarity to the earlier transition groups.

d-BLOCK ELEMENTS—GROUP VIII: IRON GROUP

Iron	Cobalt	Nickel
2.8.(2.6.6)2	(2.6.7)2	(2.6.8)2
Ruthenium	**Rhodium**	**Palladium**
2.8.18(2.6.7)1	(2.6.8)1	(2.6.10)
Osmium	**Iridium**	**Platinum**
2.8.18.32(2.6.6)2	(2.6.9)	(2.6.9)1

This group is in many ways the most difficult to treat comparatively. There have been several ways of dividing up the nine metals—vertically or horizontally. In some treatments, the vertical division has been extended to give Groups VIII, IX and X (or VIIIA, B and C), which might be thought justified by the numbers of electrons in the outer two shells of these atoms. A common arrangement is to deal with the three horizontal triads. Sidgwick adopted a third course owing something to each of these treatments: he first considered the iron triad—the ferrous metals, then divided the remaining six elements vertically into three subgroups.

These elements were given the name 'transition' before the wider application of the term in its modern significance became common. Now, in the transition elements from Group IIIA to VIIA the group number (equal to the highest valency) equals the total number of electrons in the outer two shells, less 8. Applying this formula to the above elements, we would obtain 8, 9 and 10 for the three vertical groups. The valency 8, however, is only reached by ruthenium and osmium, and the valency does not thereafter exceed 6. To continue the group numbering to 9 and 10 is therefore pointless, and if a vertical division is wanted, VIIIA, B and C is more commendable. The horizontal arrangement in three triads is not supported by the properties of the metals, which are in general no more similar than any other three adjacent transition elements along a period.

The properties which are generally found through the transition Groups IVA to VIIA show that the second element resembles the third more closely than the first element, because of the lanthanide contraction. The second and third elements show greater stability in the higher valencies, less tendency to ionize, more tendency to form complexes, and greater 'nobility'. These properties are carried forward into Group VIII, and explain why Sidgwick split up the group as he did. It is the most convenient arrangement, provided the reasons behind it are borne in mind.

Iron, cobalt and nickel differ from the platinum metals (the other six of Group VIII) in being the only ones which readily give simple ions; in rarely exceeding valency 3; and in being much more reactive. In these general points they resemble each other, as also in such transition properties as catalytic activity and coloured ions, but in detail there are differences. Iron is unique in its magnetic properties, and is far more reactive chemically than cobalt or nickel. It decomposes steam at red heat, giving Fe_3O_4 (cobalt and nickel decompose steam at higher temperatures and give lower oxides). Ferrous salts are more highly ionized than ferric, and ferrous hydroxide is a stronger base than ferric hydroxide. However, ferric hydroxide is the more insoluble, and this encourages simultaneous hydrolysis and oxidation of ferrous salts. In valency state 3, iron is amphoteric, though the ferrites are not very stable. Iron shows a greater tendency to form complexes in the ferric state than the ferrous, and on the whole the resulting complexes are more stable. The conclusion is therefore that divalent iron is more stable when ionized, and trivalent iron more stable when covalent. The only other valency (except in some carbonyls) is 6, shown in the ferrates, salts of the unknown trioxide; these are strong oxidizing agents.

Cobalt is intermediate in properties between iron and nickel, but resembles iron more closely. It is practically confined to valencies 2 and 3, the former found in all the simple (ionic) salts, which form few complexes, while the latter valency almost always gives rise to covalent complexes, as the ion is very unstable. This is an exaggeration of the behaviour of iron in these two valency states. The only higher valency established for cobalt is 4, in a small group of complexes.

Nickel is more resistant to attack by acids than the other two metals, and also resists oxidation, probably by forming an impervious layer of oxide. It differs from the other two metals in forming a few monovalent compounds, chiefly cyanides and complex cyanides. Its main valency is 2, in which it forms a number of salts. As with iron and cobalt, there is comparatively little tendency to form divalent complexes, and these are not very stable as a rule. There is some evidence of trivalency for nickel, but very few compounds are known at this valency, and there is no clear evidence of any valency higher than 3.

These three metals are outstandingly important in

GROUP VIII

engineering. Iron is the fourth most abundant element, cobalt and nickel are about a thousandfold less abundant, but good ores are available. Both elements are important constituents of special alloy steels, and nickel is also important as a plating metal, an application which takes advantage of its greater resistance to oxidation compared with iron.

Whereas in the iron triad we find a decreasing reactivity in the order iron, cobalt, nickel, the platinum metals show the reverse order, and palladium and platinum are more easily attacked than the other members of their triads. The pairs of elements taken vertically show closer similarity than neighbours taken horizontally; the six elements are only united in such general properties as their lack of reactivity to acids, as compared with most other elements, and in properties which are shared with other transition elements.

Ruthenium shows evidence of all valency states up to 8, the group valency. The metal is very resistant to attack by acids but is susceptible to oxidation. Its chief valency is 3; there are not many compounds at high valencies. Ruthenates are known (valency 6) corresponding to ferrates, and so are perruthenates (valency 7) corresponding to permanganates. Only the tetroxide represents valency 8, and this compound is unstable and an oxidizing agent.

Osmium differs in the greater emphasis on higher valencies, except that no compounds are known showing valency 5 or 7. Its main valency is 4, with a number of compounds at valencies 6 and 8. In the latter case, for instance, the tetroxide is more stable than ruthenium tetroxide, and gives rise to a number of complex derivatives as an acid. The volatile octafluoride reported previously has been shown not to exist. Both ruthenium and osmium show a strong tendency to form complex compounds, and do not form simple ionized salts.

The relationship of rhodium to iridium is not quite that of ruthenium to osmium. The main valency of rhodium is 3, and it forms few compounds at other valencies. It has a strong tendency to form complexes, like cobalt. It is a very unreactive metal with acids, and is therefore a valuable plating metal. Iridium is similarly unreactive, and has a number of special applications for this reason. Its main valency is again 3, with a series of compounds similar to those formed by rhodium. There are several compounds showing valency 4, mainly complex, and a few at valency 6 (the highest valency) including an unstable trioxide and a volatile hexafluoride. Although, therefore, iridium shows a greater stability in the higher valencies than does rhodium, it shares the same main valency with both cobalt and rhodium.

Palladium is more readily attacked by acids, halogens and oxygen than the other platinum metals. Its main valency is 2, like that of nickel, to which it has some resemblance, but it has the subsidiary valency 4, in which it forms a number of complexes; and it also forms a few trivalent compounds, mostly unstable. Even the divalent halogen compounds fail to ionize, however, and associate by coordination. Platinum resembles palladium in being comparatively readily attacked by acids, more readily than osmium or iridium. Valency 2 is still important for platinum, but valency 4 is now even more stable, and there are a few compounds showing valency 3 and a few showing valency 6. In both its main valencies many complexes are formed. Thus platinum has a number of points of similarity to palladium, and shows the differences which would be expected after considering the other elements.

GROUP VIII

The main tendencies when considering the nine elements of the whole group are the disappearance of ionization and an increase in the number of complexes formed, together with an increase in stability of the higher valencies, when passing from the first triad to the second and third. These same points may be noted with other transition groups as was mentioned above, so that no sharp division may be drawn between Group VIIA and Group VIII. Equally, the next succeeding elements in the respective periods are copper, silver and gold, which may be considered the last of the transition elements, and which share with Group VIII the unreactivity or nobility. It is important to see the Periodic Table as a whole in this way.

d-BLOCK ELEMENTS—GROUP IB: THE NOBLE METALS

    Copper 2.8.(8.10) 1
    Silver   2.8.18.(8,10) 1
    Gold   2.8.18.32.(8,10) 1

It will be noticed that the atomic structures show completed *d* shells at the expense of an *s* electron. This is because there is an 'energy bonus' when a shell is completed in this way.

The three metals, copper, silver and gold, show a fairly well marked gradation of properties, though in some respects silver is anomalous, e.g. in its valency pattern.

The noble metals have high melting points (in the region of 1000°C), are relatively soft and readily malleable and ductile; gold leaf can be beaten so thin that it transmits light. The densities are high, and rise with atomic number—8·93, 10·50, 19·32 are the three values. The atomic radii are low, that of gold is about equal to that of silver and only a little higher than that of copper. In each of their respective periods, these elements have almost the smallest atomic radii, with the exception of the Group VIII transition metals (see Appendix, p. 105). The elements are very good conductors of heat and electricity.

Their attractive polish and resistance to corrosion, coupled with their occurrence in the native state, made these metals both useful and desirable to man from the earliest times, and they have been used in most coinage systems.

GROUP IB

The metals are attacked by only those acids which are also oxidizing agents—nitric acid or hot concentrated sulphuric acid, in the case of copper and silver, while aqua regia is necessary for gold. Copper and silver become tarnished on the surface by substances containing sulphur, and copper will also slowly acquire a surface film of oxide, a process hastened, of course, by heating.

The oxides and other compounds of these metals are very readily reduced, and the metals may easily be displaced from their solutions by a base metal. Silver and gold, in finely divided form, may be recovered from their compounds simply by heating.

The metals can all show the valency 1. In the case of copper, the more stable valency is 2, and the monovalent substances are either insoluble or complex. In the case of silver, valency 1 is stable, and only the strongest oxidizing agents will compel it to exert a valency of 2. With gold, the stable valency state is 3, and it resembles cuprous copper in valency state 1. In each case the monovalent metal compounds are colourless, while the higher-valent compounds are coloured. Monovalent copper (cuprous) halides resemble the silver halides in being insoluble in water but soluble in ammonia. Monovalent gold halides are unstable.

In addition to variable valency and some coloured compounds, the metals have the transition property of catalytic activity. Their electronic structure has in each case a penultimate shell of 18, with a single electron in the outer shell. This single electron is insufficient to stabilize the group of 18, which needs at least two electrons in the outer shell (with, of course, a correspondingly higher positive charge on the nucleus). There is very little energy difference between the orbital of the 18th electron of the penultimate shell and the next available orbital of the outer shell. It is therefore possible for one, or two, extra electrons to become available for valency purposes, so that these metals are the last to show transition properties in each of their periods. The end of the transition series depends on the definition employed; these elements may be regarded as transitional in their higher valency states.

The single readily available electron accounts for the good electrical conductivity of these metals. The electronic structure also accounts for the chemical properties, for an element is only highly reactive if there is a much more stable structure readily attained by reaction. In this case, no possible attainable ionic structure is stable. The small ionic charge is offset by the instability of 18 as an outer group, and by the small size (low atomic volume) so that the ions attract electrons powerfully, and the change $M^+ \to M$ is easy.

The ionization potential for copper is about twice the value for caesium, and there is an increase from Cu to Au, in place of the decrease from Li to Cs. On the other hand, the silver ion, Ag^+, is not apparently much hydrated, judging by its mobility; it is impossible to investigate Cu^+ or Au^+, and Cu^{++} (which is much hydrated) is not directly comparable.

The metals readily form covalent complexes, in which they attain a much more stable structure with the shared electrons, usually totalling 8. The ability of copper and silver to coordinate with ammonia is well known, and made use of in analysis. Similarly, all the metals will coordinate with cyanide and thiosulphate radicals. While copper and silver do form some simple electrovalent compounds, it appears that all gold compounds are covalent, and frequently complex.

In this subgroup there is therefore a decreasing

electropositive tendency with increasing atomic number, unlike the position with the alkali metals. The explanation is based upon the unusually small atomic volume of silver and, particularly, gold; this in turn is part of a more general phenomenon, the lanthanide contraction. In the subgroups of transition elements from VA to VIII, immediately preceding the noble metals, we find a similar decrease of electropositive nature with increasing atomic number, so that these noble metals show their transition properties in this respect also.

d-BLOCK ELEMENTS—GROUP IIB: ZINC, CADMIUM and MERCURY

Zinc   2.8.(8,10).2
Cadmium 2.8.18.(8,10).2
Mercury 2.8.18.32.(8.10).2

This group concludes the *d* block, and in each case the *d* orbitals are filled. The extra positive charge on the nucleus compared with the preceding group ensures that none of the *d* electrons can be removed by oxidation, and these elements have none of the transition properties caused by incomplete *d* orbitals in the rest of the *d* block.

The metals do not show the smooth gradation of properties shown by some previous groups. This is mainly because mercury has some highly individual properties which separate it from zinc and cadmium; of all the metals the diagonally-related silver is the element to which mercury shows the closest similarity. Zinc and cadmium show a number of points of similarity to magnesium and calcium, as well as to beryllium.

Group IIB elements decrease in electropositive tendencies with increasing atomic number, although the change is not marked between zinc and cadmium. These two metals oxidize only superficially in air at room temperature, and are used as plating metals. They will, however, oxidize readily at elevated temperatures. Zinc normally dissolves readily in dilute acids, but pure samples are much more resistant to attack than commercial metal; cadmium is more resistant still. Mercury is only slowly attacked by *oxidizing* acids, and it is very easily displaced from its compounds by reducing agents, or even by heating, as with silver and gold.

Zinc oxide and hydroxide are amphoteric (which is why zinc dissolves in caustic alkali), and this is an important point of difference from magnesium. Cadmium oxide is basic, and the hydroxide is probably insoluble in excess caustic alkali, although this is strongly adsorbed by the precipitated hydroxide. Mercuric oxide is basic and shows no acidic properties at all. The three oxides, and also the sulphides, show the same covalent lattice structure.

The elements are uniformly divalent; even mercury in the mercurous compounds shows the valency 2. The behaviour of the compounds as electrolytes is, however, not quite straightforward. Zinc and cadmium salts ionize to a considerable extent, but the evidence is conflicting and incomplete. Comparing the halides, the two fluorides are both non-volatile. Zinc chloride, bromide and iodide melt below 500°C as does cadmium iodide; they are all appreciably soluble in organic solvents (as well as in water). The aqueous solutions conduct electricity, but this may be due, at least in part, to hydrolysis. Cadmium bromide and chloride, on the other hand, melt above 500°C, and are not very soluble in organic solvents. (A melting point above 500°C for a chloride is normally taken to indicate electrovalency, and below 500°C, covalency.) Conductivity measurements indicate that the principal ionization is of the type:

$$CdCl_2 \rightleftharpoons CdCl^+ + Cl^-$$

with rather less of the complete ionization, and some tendency to complex formation.

Mercury compounds, for the most part, do not ionize appreciably, the exceptions being mercury fluoride and the salts of the strong oxyacids, nitric and perchloric acids.

On examining the degree of hydration of various salts, it is found that zinc salts are nearly aways highly hydrated and are hygroscopic, while cadmium salts, on the average, are slightly less hydrated and show less tendency to be hygroscopic. Mercury compounds differ sharply by being usually anhydrous (one point of resemblance to silver), except for those few ionized salts of oxy-acids where the anion can be hydrated to the extent of 1 or 2 H_2O.

All the metals readily form complexes, owing to their tendency to assume the covalent state. Zinc and cadmium form many compounds involving direct links to carbon, as well as nitrogen, sulphur, etc., but zinc appears to form a more stable link with oxygen than does cadmium. Mercury is outstanding for the stability of its link with carbon and the weakness of the link with oxygen; it again forms complexes involving nitrogen and sulphur, and also halogen.

Mercury differs from the other elements in a number of important respects, some of which have been mentioned. It is more volatile than any other metal, having an appreciable vapour pressure at room temperature, and boiling at 357°C to give a monatomic vapour. There are few metals boiling below 1000°C, but both zinc and cadmium are among them, boiling at 906° and 764° respectively. The comparative unreactivity of mercury is paralleled by its very high ionization potential and electrode potential. The metal is readily discharged from its compounds by reducing agents. Mercuric fluoride is sparingly soluble and hydrolysed; the chloride is appreciably soluble, too, but the bromide and iodide are almost insoluble. In these solubilities mercury shows a stronger similarity to silver than to cadmium or zinc. Mercury shows no tendency to complex formation to increase the covalency beyond 2, being apparently satisfied with only 4 electrons shared; this is without parallel, and not easily explained. The refusal to form an oxy-acid anion is in line with the wholly-basic (i.e. not amphoteric) nature of mercuric oxide, but out of character with the otherwise feebly electropositive behaviour of mercury. The apparent monovalency in the mercurous compounds, in which we have one link between two mercury atoms, $(Hg-Hg)^{2+}$ is also unusual and difficult to understand. The other features of its behaviour suggest that in this element we have the first example of the 'inert pair' behaviour, as found with tin and lead, for example (see pp. 15 and 30). As mercury only has two electrons in the outer shell this means that it behaves in some ways like an inert gas—for instance, in readily giving a monatomic vapour at a moderate temperature.

The heavier elements in each of the p block groups show this tendency for a pair of electrons to become inert, and in this property we see a horizontal rather than a vertical relationship for mercury.

f-BLOCK ELEMENTS: LANTHANONS AND ACTINONS

	(57) Lanthanum	2.8.18.(2,6,10)(2,6,1) 2
	(58) Cerium	2.8.18.(2,6,10,1)(2,6,1) 2
to	(71) Lutetium	2.8.18.(2,6,10,14)(2,6,1) 2
	(89) Actinum	2.8.18.32.(2.6.10)(2,6,1) 2
	(90) Thorium	2.8.18.32.(2,6,10,1)(2,6,1) 2
to	(103) Lawrencium	2.8.18.32.(2,6,10,14)(2,6,1) 2

The two f block series consist of the lanthanons, sometimes called lanthanides, and the actinons (actinides). The former have atoms in which the $4f$ shell is building up, and the latter have atoms in which the $5f$ shell is building up. Lanthanum and actinium, which give the series their names, have no f electrons and can be found in the d block. However, as can be seen from the structures given above, the two name elements have the same electronic structures in their outer two shells that are found in all the f block elements. Because the outer two shells are constant, these f block elements have long been known as 'inner transition elements'; whereas transition elements vary in their penultimate shells, inner transition elements vary in their ante-penultimate shells.

The constancy of the two outer shells results in exceptionally close similarity in chemical properties, particularly of the lanthanons, but also of the later actinons.

The predominant valency throughout the lanthanons is 3, making use of the 2 s electrons and the 1 d electron. Only cerium exhibits a stable valency of 4, although this element also exhibits valency 3. As a block, the lanthanons show catalytic activity and some ions are coloured; three or four elements show variable valency to 2 or 4 while keeping the main valency 3. The metals form strongly basic oxides and are quite highly electropositive; there is only a little tendency to form complexes.

Thus most of the 'transition properties' are represented among them (see also the chapter on transition elements). The atoms, and resulting ions, decrease steadily in size with increasing atomic number (the 'Lanthanide contraction'; see *Figure 5*) and with the

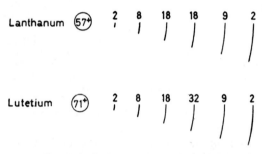

Figure 5. The 4th shell of electrons does not require more room for 32 than for 18 electrons, hence the 5th and 6th shells are under the influence of 14 additional positive charges in Lu, an increase of approx. 25 per cent

decreasing ionic radius there is a steady decrease of electropositive nature, shown by a decreasing basic strength and an increasing stability of those complexes that are formed. This fact is employed during purification by ion exchange; a solution of mixed lanthanide ions is passed through an ion exchange column, and then the column is eluted by means of a complexing agent. The lanthanons are eluted in reverse order of atomic number.

Outside the lanthanons, the effect of the lanthanide contraction is to make the atoms of the later transition elements smaller than would otherwise be expected; this makes the elements less electropositive, more noble, than the earlier transition elements (see *Figure 6*).

III	IV	V	VI	VII	VIIIA	B	C	
Sc	Ti	V	Cr	Mn	Fe	Co	Ni	Cu
Y	Zr	Nb	Mo	Tc	Ru	Rh	Pd	Ag
La–Lu	Hf	Ta	W	Re	Os	Ir	Pt	Au
Ac –								

Figure 6. In Groups IV–VIII there is a greater difference between the 1st and 2nd elements than between the 2nd and 3rd; the latter gap gradually increases. The effect of the lanthanide contraction is still noticeable in VIIIC but has disappeared in IB

Of the second series only actinium, thorium, protoactinium and uranium are known in nature, and the position of these elements in the Periodic Table was open to debate before the heavier elements, known collectively as transuranic elements, were synthesized. The discovery of the transuranic elements by Seaborg and his co-workers in the period since 1940 has been of the greatest importance, and has led to a considerable revival of interest in inorganic chemistry. There are many reasons for this importance, both theoretical and practical, as will be seen.

The fact that these elements are synthetic means the end of the alchemists' search for transmutation. The elements are made atom by atom, by bombardment with neutrons or accelerated positive ions. In the first case, when neutrons are absorbed by ^{238}U, the product atom ^{239}U is a β-emitter, giving therefore element number 93 (neptunium) (p. 100). This in turn is a β-emitter, and gives element number 94 (plutonium, ^{239}Pu). The process becomes possible on a large scale by employing the intense neutron fluxes in atomic reactors. Plutonium is of military importance, and will no doubt prove of commercial value as an atomic fuel, because its atoms are fissile in a stream of slow neutrons, with the release of energy and more neutrons.

The method of neutron bombardment is useful for elements just above uranium, but for elements several places above, it is necessary to use highly accelerated positive ions, such as alpha particles, and also 'stripped' atoms such as carbon, with all electrons removed. These nuclear reactions become more and more difficult to bring about as the atomic number increases, because of the potential barrier round the nucleus. The result is that yields progressively decrease as atomic number increases; in the case of element number 102 (nobelium), made in 1957, a total of 17 atoms were observed. This was enough to assign a half-life of about ten minutes.

Among the transuranic elements now known are members of the $(4n+1)$ radioactive series, previously unknown. This series includes, as daughter products, isotopes of francium and astatine (87, 85) which are by-passed by other series, and it finishes with a 'stable' isotope of bismuth, instead of lead.

Since these elements are produced, at any rate in the first instance, in very tiny quantities, new and elegant techniques have been developed for working on a smaller scale than ever before. Recognition of the element is helped by characteristic radioactivities, but even so the difficulties are very great. However, the chemistries of many of the transuranic elements have now been worked out in detail; the techniques involved are of great practical significance, and are having an effect in many other spheres of work. They have already been employed in a re-examination of some radioactive elements such as polonium, whose chemistry presented many difficulties and was therefore only sketchily known. This element, previously isolated in tiny amounts from pitchblende, may now be obtained by bombardment of bismuth, which makes it a little more accessible.

The examination of the transuranic elements finally resolved the controversy about the fourth long period of the Periodic Table. This period is incomplete, and after the electron structure of the lanthanons had been settled, the question arose whether there would be a second 'inner transition' series, and if so where it began. It is now known that the lanthanon series is repeated in the same group, starting with actinium, number 89. Thorium (90) is a close analogue of cerium, and so on; as each transuranic element was made it was found to be analogous to the corresponding lanthanon. This is perhaps shown most clearly in the separation by means of ion exchange, when the resemblances are so consistent that appearance in the expected drop of eluate can almost be used as confirmation of identity of a new element.

A consideration of the first few members of the actinon series quickly shows that they are more easily oxidized than the corresponding lanthanons, and to a greater extent in some cases. Thorium shows no trivalent state to correspond with cerous compounds; uranium shows a stable valency of 6, unlike neodymium, and so on. This general tendency is continued with neptunium and plutonium, and it is not until curium (analogue of gadolinium) is reached that we find a stable maximum valency of three. The readier oxidation must mean that the energy levels are correspondingly close together; this is the general trend in the later periods, as shown by other evidence. In studying the normal transition series, it was noted that the heavier elements in each group showed a greater stability in higher valencies, and were less easily reduced in their highest valency state. The main difference between the actinons and lanthanons is therefore to be expected from other observations.

As a result of the production of plutonium on a large scale, several of the transuranic elements are likely to become available as by-products also on a fairly large scale. This means that they could be employed for their characteristic radiation; americium for instance can be used for radiography.

ELECTROPOSITIVITY AND ELECTRONEGATIVITY

ELECTROPOSITIVITY

The electropositive elements are the metals, and the most typically metallic elements, speaking chemically, are those which are most electropositive, i.e. have the strongest tendency to lose electrons: $M \rightarrow M^+$. This tendency to lose electrons may be measured by the electrode potential, but in solution there may be complicating factors such as the hydration of the ions. The gaseous ionization potential can, however, be measured spectroscopically, without complications due to hydration. Elements can be arranged in order of ionization potentials in the form of the electrochemical series; the lower the ionization potential (in electron volts) the smaller the energy required to remove the electron, and the more electropositive the metal. It should not be forgotten that a number of metals are less electropositive than hydrogen; these are metals which do not release hydrogen from acids.

Although, for some purposes, electropositivity has been replaced by electronegativity, with a quantitative significance, it is convenient still to use the former term when speaking about metals, which, after all, are in the majority among the elements. It is necessary to have an understanding of the chemical properties which belong most typically to electropositive elements; an examination of a family of elements reveals a gradation of these properties.

The tendency for atoms to lose electrons is governed by Fajans' rules, and is greatest for large atoms

ELECTROPOSITIVITY

which give monovalent cations. In large atoms, the outermost (valency) electrons are in a weaker field than in small atoms, being further from the positive nucleus. Each successive electron to be removed will need a greater force than the previous one, because of the positive charge to be overcome. Therefore, electropositive behaviour becomes less marked as the size of the atom decreases, and as the value of the charge on the resultant cation increases.

Those elements which are highly electropositive give compounds which are strongly ionized in water or in the fused state, and are non-volatile. Oxides and hydroxides are basic, because of the ionization: $MOH \rightarrow M^+ + OH^-$, rather than $MOH \rightarrow MO^- + H^+$ (the alternative for electronegative elements). There is a marked general reactivity with acids and, in extreme cases, with water. As the less strongly electropositive elements are reached, the use of oxidizing agents is necessary before the element will react. Only those elements which are strongly electropositive will, in general, form salts with oxyacids, such as carbonates and sulphates; the sulphate or carbonate radical is not as powerful as, say, a fluorine atom in removing electrons from atoms to form cations.

Just as we may compare the ease of removal of an electron $M \rightarrow M^+$ in judging electropositive behaviour, so we may compare the difficulty of replacing that electron. There is one common process which tends to bring about the reverse change, $M^+ \rightarrow M$ and that is hydration of the cation by donor action of oxygen: $H_2O \rightarrow M^+$, involving one or more water molecules. The positive charge is spread over the whole complex, and is no longer exclusively found on the metal. The more readily the ionization $M \rightarrow M^+$ proceeds, the less easily will the reverse change take place, and therefore the less hydrated the cation will be. Conversely, if the ionization is not strong, the cation will be more ready to accept hydration. By comparing the average degree of hydration of salts, therefore, a series of metals may be readily arranged in order of electropositivity. Anionic hydration is of much less importance than hydration of cations, but its effect can be eliminated by comparing salts of the same anion.

The salts of weakly electropositive metals hydrolyse more readily than those of 'strong' metals; there may be oxy-salts or basic salts formed. Again, weakly electropositive metals will tend to form more complexes than strong metals, as the tendency to form covalent complexes is approximately inversely proportional to the degree of ionization.

We may, therefore, draw up a scale for each of the seven main chemical properties, which may then be used to compare the electropositive behaviour of families of elements: degree of ionization of comparable compounds; relatively basic or amphoteric nature of oxides or hydroxides; reactivity with acids; number of salts formed with oxyacids; extent of hydration of salts; degree of hydrolysis of salts; and the relative number of complexes formed. It does not follow that the same indication will be given by all these properties, nor that the order based on chemical properties will be identical with that based on physical properties, although there is general agreement.

When the groups of the Periodic Table are compared, it is found that in some cases elements show an increase in electropositive character with increasing atomic number, but in other groups the reverse is true. The whole picture may best be seen in *Figure 7*.

It will be noted that the 'extreme' groups all show

the first pattern, while the 'central' groups show the reverse. In each case there is some uncertainty, due to conflicting evidence, where the changeover occurs: this, it will be noted, is only to be expected from general considerations. Those 'central' groups in

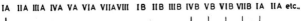

Figure 7. The arrows point in the direction of increasing electropositive character. Dotted arrows indicate conflicting evidence or the absence of a marked trend

which there is a decrease of electropositive character with increasing atomic number are the groups in which there is little or no increase in atomic size with atomic number; atoms which are compact and heavy, although metallic, do not produce simple cations and are 'noble' in behaviour.

ELECTRONEGATIVITY

On p. 86 it was mentioned that electronegativity has been given a quantitative significance. This was first done by L. Pauling (1932 onwards). He derived his values originally from considerations of bond energy, but the figures themselves are arbitrary. They can be used to indicate relative bond energies and stabilities of compounds. The most highly ionized compounds result from combination of elements which differ most in the values of electronegativities, and covalent compounds are formed when elements of similar electronegativities unite.

The following table shows some values for the elements arranged as in the Periodic Table.

H						H
2·1						2·1
Li	Be	B	C	N	O	F
1·0	1·5	2·0	2·5	3·0	3·5	4·0
Na	Mg	Al	Si	P	S	Cl
0·9	1·2	1·5	1·8	2·1	2·5	3·0
K	Ca				Se	Br
0·8	1·0				2·4	2·8
Rb	Sr				Te	I
0·8	1·0				2·1	2·4
Cs	Ba					
0·7	0·9					

It will be seen that, in the groups shown, the electronegativity decreases with increasing atomic number, in agreement with the increasing electropositive character of the elements. Lines of equal electronegativities would be diagonal lines across the table, in the same sense as the diagonal relationships previously commented on (p. 11).

TRANSITION ELEMENTS

The first two periods of the Periodic Table, called 'short periods' each contain eight elements, which show a fairly regular gradation of properties as the outer electron shell builds up. Elements range from markedly electropositive alkali metals to markedly electronegative halogens. The long periods, on the other hand, contain a number of neighbouring elements with strong resemblances to each other, as well as elements corresponding to those in the short periods. These elements, showing similarities to their 'horizontal' neighbours and thus cutting across earlier group resemblances, are termed 'transition elements', and are all metals. It is difficult to define the transition state rigorously in a few words.

The electronic structure of the atoms provides the explanation for the transition state, as outlined in Chapter 1. In the two short periods, each element has one more electron in its outer shell than its predecessor. The eight elements from lithium to neon have from one to eight electrons in the second shell, and the eight from sodium to argon repeat this pattern in the third shell. This third shell is temporarily full with eight electrons, and potassium and calcium begin to build up the fourth shell, as indicated (on p. 9). From scandium onwards, extra electrons build up the third shell to its eventual total of 18, entering the $3d$ orbitals. There are 5 orbitals which can accommodate a maximum of 10 electrons.

These elements therefore appear in the d block in the Periodic Table. They are paralleled by elements 39–48 in which the $4d$ shell is being filled. However,

THE PERIODIC TABLE

			3d					4s
21	Scandium	(3s + 3p – filled)	⊕	◯	◯	◯	◯	⊕⊕
22	Titanium		⊕	⊕	◯	◯	◯	⊕⊕
23	Vanadium		⊕	⊕	⊕	◯	◯	⊕⊕
24	Chromium		⊕	⊕	⊕	⊕	⊕	⊕
25	Manganese		⊕	⊕	⊕	⊕	⊕	⊕⊕
26	Iron		⊕⊕	⊕	⊕	⊕	⊕	⊕⊕
27	Cobalt		⊕⊕	⊕⊕	⊕	⊕	⊕	⊕⊕
28	Nickel		⊕⊕	⊕⊕	⊕⊕	⊕	⊕	⊕⊕
29	Copper		⊕⊕	⊕⊕	⊕⊕	⊕⊕	⊕⊕	⊕
30	Zinc		⊕⊕	⊕⊕	⊕⊕	⊕⊕	⊕⊕	⊕⊕

those chemical properties which have been observed to belong to the majority of transition elements, have been shown to depend on the possibility of an *incompletely-filled* set of orbitals (other than those in an outer shell). It is therefore not surprising that transition properties are not found in zinc or the zinc group elements, in which the *d* orbitals are filled, as well as the *s* orbitals of the next shell. Transition elements should not therefore be defined as '*d* block' elements —this is too simple. It should be remembered that the *p* block concludes with the noble gases, having a completely-filled set of *p* orbitals, and that noble gases have properties different from those of the other *p* block elements.

The cases of chromium and copper need further comment. When the chromium structure is compared with that of vanadium, it will be observed that an electron has been 'pulled back' from the *4s* orbital so that the *3d* orbitals each contain one electron. From this type of observation, Hund formulated his 'maximum multiplicity' rule, which states that the ground state of an atom contains the maximum number of unpaired electrons. This is the state with least energy for the particular atom. The ground state of copper again shows an electron 'pulled back' from the $4s$ orbital, but this time in order to complete the filling of the $3d$ orbitals. This is not incompatible with Hund's rule quoted, for each possible structure contains one unpaired electron. It will be noted that it was stated above that transition properties depend on the possibility of an incompletely-filled set of inner orbitals. In the case of copper, it is found that very little energy is necessary to 'promote' one of the $3d$ electrons to the $4s$ level—this corresponds, indeed, to the cupric state, in which there is present a vacant inner orbital. One is justified in including copper (and the copper group) under the heading of transition metals.

One of the more significant properties of transition elements is that of variable valency. Iron shows valencies of two, three, and six; manganese can have valencies from two to seven. This is now seen to be due to the small difference in energy levels between orbitals in the outer and penultimate shell (in the case of the first long period, the $4p$ and $3d$ orbitals). The chemical energy of oxidation is sufficient to raise the $3d$ electrons to the vacant $4p$ orbitals, thus making extra electrons available for valency purposes*. The neutral manganese atom has the structure $2.8.(8.5)2$, and the manganous ion is $2.8.(8.5)^{2+}$. This is not an inert gas structure, and the five $3d$ electrons can be removed by successive oxidation. The higher valencies are of course covalencies, not electrovalencies, and

* Hybridization of the valency electrons among the various available orbitals then takes place, ensuring equivalence.

manganese readily forms the symmetrical permanganate ion with a single negative charge. In accordance with Fajans' rules, covalency formation is encouraged by the small atomic volume of manganese and the other transition elements. They are found at the minima of Lothar Meyer's curves, and there is not as great an increase in atomic volume in descending the transition groups as there is in other groups.

Transition metals are active as catalysts, either in the metallic state or in the form of compounds. One theory of catalytic activity is based on the formation of intermediate compounds, and it is clear that these metals with their variable valency will be particularly susceptible to the formation of unstable intermediates.

Compounds of the transition metals, whether ionic or covalent, are usually markedly coloured, whereas compounds of other elements are not often coloured. Colour is caused by some visible wavelengths of incident light being absorbed preferentially. This occurs when the quantum energy of light of a particular wavelength corresponds to the difference in energy level between two possible orbitals, and is again due to the small energy difference between the outer shell and penultimate shell. When the energy differences are larger, ultra-violet light may be absorbed, but not visible light. Absorption of light energy gives an 'excited' atom, in which some electrons have been promoted to higher energy levels.

Metals of the transition series are paramagnetic, that is, tend to move in the direction of greater density of the lines of magnetic force. This property is partly due to electrons with unpaired spin, and these elements with incomplete electron shells have in most cases several such electrons. Transition metal compounds may also be paramagnetic, and a study of

TRANSITION ELEMENTS

magnetic properties provides useful information on electronic structure.

The simple ions of the transition elements, as was pointed out above, do not have the structures of inert gases, and are not usually very stable under oxidizing conditions. Intermediate oxidation states in simple covalent compounds are not, however, necessarily more stable (e.g. covalent Mn^{IV} is 2.8.(8.3.)8). They may be stabilized by complex formation, and transition metals are particularly likely to form complexes. This property is not confined to transition elements so exclusively as the other properties mentioned earlier; it is found with many of the metals which are not highly electropositive.

The structure of the transition elements was explained above in terms of the first long period. The second long period is similarly constituted, with the $5s$ orbital filled first, followed by the $4d$ orbitals. The last of the $4d$ orbitals is filled in the case of silver, which has only one $5s$ electron in the ground state, and is consequently predominantly monovalent: $5s$ (2); $4d$ (10); $5p$ (6), again a total of 18. The earlier transition elements of this period (e.g. Zr, Nb, Mo) are more easily oxidized to the group valency than the corresponding metals of the first long period, due to more easy removal of the $4d$ electrons from an incomplete shell compared with the $3d$ electrons in the first long period. In silver, the newly-completed shell resists attack by any but the strongest agent (fluorine). In this it differs sharply from copper, which is much more easily oxidized, as well as from the earlier metals of its period.

The second long period is complete with the rare gas xenon, structure 2.8.18.18.8. But just as the third shell could expand from 8 to 18 when the $4s$ orbitals had been filled, so the fourth shell can eventually expand to 32; this expansion begins when the *6s* orbitals are filled, as the *4f* orbitals are higher in energy level than the *6s* but lower than the *5d* or *6p*. Because there are two electron shells containing some electrons outside the fourth shell, the series of elements in which the expansion from 18 to 32 occurs may be termed 'inner transition' elements; they are known as the lanthanons, and are exceptionally similar to each other.

The third long period then continues with the expansion of the fifth shell: the *5d* orbitals fill up in the elements from hafnium to gold: *6s* (2); *4f* (14); *5d* (10); *6p* (6), a total of 32. The fourth long period is incomplete, but closely echoes the third. Actinium begins the second series of 'inner transition' elements, known as the actinons, in which the *5f* orbitals are successively filled. This series includes the transuranic elements up to number 103, and much of the interest in these elements centres round the problem of their atomic structure compared with the structure of the lanthanons.

Transition elements are a normal feature of the Periodic Table, recognized since the quantum treatment of electronic shell structure. Their successful explanation in terms of electronic structure gave extra confidence in electronic theory and also strengthened the value of the periodic classification.

OCCURRENCE OF THE ELEMENTS

There are several factors which determine the way in which the elements occur as main ores, and by taking account of these factors it is possible to see a certain order in what otherwise seems arbitrary. The main factors are: the relative abundance of the element, its reactivity, solubility relationships, volatility, and relative sizes of atoms.

The relative abundance of the elements has been estimated by V. Goldschmidt, among other workers. Certain inferences can be drawn from a comparison of the abundances of the elements plotted against atomic number, and these have an important bearing on the stability of the atomic nuclei concerned (stability, that is, under physical forces such as bombardment with accelerated particles). It becomes plain that, in general, even-numbered elements are more abundant than odd, and heavier elements are less abundant than light. There are a few exceptions; for instance, lithium, beryllium and boron are scarcer than would be expected from this rule, and aluminium is an odd-numbered element that is among the very abundant ones (third in the list). Since oxygen is easily the most abundant element, it is clear that there is a very high chance of finding a particular element combined with oxygen, and the general principles will be clearer in this case than in that of, say, iodine, which is quite scarce.

The reactivity of the elements can be compared by means of the position in the Electrochemical Series, or by the electronegativity, a measure of the electron-attraction of an atom. It is found that, in general, the most electropositive elements are found united to the most electronegative ones, and the next most electropositive elements are united to the most electronegative elements still left. The alkali metals occur as halides, chiefly chlorides; the alkaline earth metals united with oxygen, and the highly basic oxides have then united with acidic gases, to form carbonates and sulphates. Then follow metals which have remained as oxides, being slightly less basic, and the same metals also tend to occur as sulphides. Only a few 'noble' metals occur native to any extent.

This factor alone cannot account for the occurrence of a number of ores; for instance, it does not explain why fluorine occurs largely as calcium fluoride and not, say, potassium fluoride. (It does also occur as sodium fluoride jointly with aluminium fluoride in cryolite.) Solubility in water is a factor accounting for the chlorides, bromides and iodides found in sea water and for the existence of some deposits from dried-up seas, such as Cheshire salt. It may be imagined that fluorides of the alkali metals could also have been present in the sea; however, many fluorides are insoluble, and of the electropositive metals which give insoluble fluorides calcium is probably the most abundant. (The heat of formation of a fluoride of a highly electropositive metal such as calcium would be greater than that of a 'weaker' metal.) Mutual solubility of compounds in the molten state has also been important at an earlier stage in the earth's history, and may explain many cases of double compounds and co-occurrence generally.

Volatility accounts for the gases of the atmosphere, for the escape of helium and hydrogen (which is very much more abundant on a cosmic scale than on earth), and for the widespread occurrence of carbonates and sulphates formed by the combination of the corresponding acid gases with oxides. It is largely responsible for the occurrence of native sulphur due to the interaction of the two gases, sulphur dioxide and hydrogen sulphide. The non-volatility of silicates and phosphates (and also their insolubility) is probably also important in preventing much exchange.

The relative sizes of atoms have an effect on the particular crystal lattice adopted, and hence on the existence or absence of isomorphism. Elements of the same family do not always occur together; if corresponding compounds cannot be isomorphous it is unlikely that they will be found in the same deposit. Similarity of size explains some cases of joint occurrence of elements which are not of the same family; thus transition metals have very similar atomic radii, and often several of them are found in one ore. The case of the rare earths is an example: here, outstanding similarity of properties is coupled with slowly changing size, and in appropriate ores a number of these elements are always present together. These ores also frequently contain yttrium, thorium and several other transition metals which may or may not correspond closely in properties but are closely similar in size.

RADIOACTIVITY AND NUCLEAR STABILITY

A feature of the table of elements is the group of heavy elements which are radioactive. There are hardly any naturally occurring radioactive isotopes of elements lighter than lead, number 82 (^{40}K is an exception), while all heavier elements are radioactive (with the possible exception of bismuth, 83). The problems posed by the discovery of radioactivity have been solved mainly during the present century, and their elucidation has helped greatly in the development of atomic theory.

Each radioactive element (nuclide is a better term) disintegrates at its own definite rate such that a given proportion of the number of atoms present disintegrates in unit time; half the atoms decay in a period known as the half life. The disintegration* is marked by the emission of either an α- or a β-particle from the nucleus: the former is a helium nucleus (two protons + two neutrons, $^{4}_{2}$He), while the latter is an electron arising from the change:

neutron $\longrightarrow$ proton + electron, $^{1}_{0}$n $\longrightarrow$ $^{1}_{+1}$p + $^{0}_{-1}$e.

Russell, Soddy and Fajans formulated their displacement law to express the effect of disintegration on the remainder-atom. Emission of an α-particle means that the nucleus left will be 4 mass units lighter, and 2 places lower in atomic number, than the atom which has disintegrated. Emission of a β-particle, on the other hand, results in an atom of the same mass

* For other, rarer, types of disintegration see specialist books.

number but of atomic number higher by one. It follows that, if an atom emits first an α-particle and then two β-particles, the product will be an isotope of the first element, 4 mass units lighter; this was the first recognized example of isotopy.

In searching for the reason why some elements disintegrate spontaneously, scientists have compared the known stable isotopes with the radioactive nuclides and have deduced the laws for nuclear stability.

(1) Neutron–proton ratio. This is never less than one, except in the case of hydrogen, which has no neutrons at all. Elements 2 to 20 have a ratio of about 1, while for heavier elements the ratio increases progressively, up to a value of about 1·5 for lead, number 82.

(2) Even number rule. Even-numbered elements are usually more abundant than the odd-numbered ones, and have more isotopes. (The numbers of isotopes of non-radioactive elements are about 220:60 for even:odd elements). Nuclides in which the number of neutrons is even also outnumber those in which it is odd, and almost all the very common elements are of the 'even–even' class; oxygen, with 8 protons and 8 neutrons, silicon $(14+14)$, calcium $(20+20)$, iron $(26+30)$ and magnesium $(12+12)$ are examples. Those nuclides in which both numbers are odd are few in number and scarce. They comprise deuterium, 1 neutron + 1 proton (the scarce isotope of hydrogen), lithium-6 $(3+3)$; boron-10 $(5+5)$; and nitrogen-14 $(7+7)$, the latter being the only reasonably abundant representative of this class.

(3) Total positive charge. It is found on inspection that the two previous rules apply also to the radioactive elements, and are therefore general rules for the existence of any nuclide. The only difference between radioactive and non-radioactive nuclides is apparently

the total positive charge; when this charge exceeds 82, the repulsion between the protons produces instability of the nucleus, regardless of the number of neutrons present. Those radioactive elements with a slightly low neutron–proton ratio are then α-emitters, while those with a higher neutron-proton ratio are β-emitters, since the result of α-emission is to raise the ratio, while β-emission results in a lower ratio.

Among 'synthetic' radioelements β-emission is the most common form of activity, since these usually have a high neutron–proton ratio, whether they are produced by absorption of neutrons in a reactor, or extracted from fission products.

Both α- and β-emission can be accompanied by γ-radiation, which is a form of energy. γ-radiation for nuclear reactions replaces the heat produced in the course of exothermic reactions, and spontaneous nuclear disintegration can be compared with an exothermic reaction, since a supply of energy is not needed to start the process.

THE CHLORIDES OF SOME ELEMENTS

Melting points and (in brackets) equivalent conductivity in fused state

LiCl 610° (166)	BeCl$_2$ 404° (0·086)	BCl$_3$ −107° (0)	CCl$_4$ −23° (0)	—
NaCl 808° (133)	MgCl$_2$ 715° (29)	AlCl$_3$ (subl.) 183° ($1·5 \times 10^{-5}$)	SiCl −70° (0)	PCl$_5$ 148° 0
KCl 772° (103)	CaCl$_2$ 782° (52)	ScCl$_3$ 960° (15)	TiCl$_4$ −23° (0)	—
RbCl 717° (78)	SrCl$_2$ 875° (56)	YCl$_3$ 700° (9·5)	ZrCl$_4$ 335° (subl.) (−)	NbCl$_5$ 194° (2×10^{-7})
CsCl 645° (67)	BaCl$_2$ 960° (65)	LaCl$_3$ 870° (29)	HfCl$_4$ 317° (subl.) (−)	TaCl$_5$ 211° (3×10^{-7})

The figure 500°C for the melting point of the chloride is taken as the dividing point between electrovalent (above) and covalent (below); that this is valid, is illustrated by the conductivities in the fused state. The line divides the two classes of chlorides; its general diagonal trend is a natural consequence of the two Fajans' rules.

TABLES OF PHYSICAL CONSTANTS

Group I (A)

	Li	Na	K	Rb	Cs
Atomic number	3	11	19	37	55
Atomic weight	6·94	22·992	39·102	85·48	132·92
Atomic structure	2, 1	2, 8, 1	2, 8, 8, 1	2, 8, 18, 8, 1	2, 8, 18, 18, 8, 1
Density (g/cm^3)	0·53	0·97	0·86	1·53	1·90
Melting point °C	180	98	63	39	28·5
Boiling point °C	1330	890	760	700	685
Atomic volume	13	23·5	45·4	55·8	70
Atomic radius Å	1·34	1·54	1·96	2·11	2·25
Ionic radius Å	0·6	0·95	1·33	1·48	1·69
Ionic mobility	33·5	43·4	64·6	67·3	68
Salts hydrated, per cent	76	74	23	3	3
Gaseous ionization potential	5·36	5·14	4·34	4·18	3·89

The figures of various physical properties are important not so much for their own sake as in showing the trends through a group and the comparison between an element and its neighbours. Figures obtained from different sources do, in fact, show slight variations from those quoted.

The figures for melting and boiling points show an orderly progression, decreasing from Li to Cs as the size of the atom increases; the interatomic forces must be decreasing with size. Atomic volume, atomic and ionic radii all illustrate the increase in size from Li to Cs. With this increase in size of monovalent metals, we have the Fajans Rule concerning size alone, operating, and the increasing ease of the change $M \rightarrow M^+$ is illustrated by the gaseous ionization potential. However, the hydration energy decreases with increasing ionic size, and this is indicated by the decreasing number of salts found to be hydrated (out of some 30 common salts).

The Li$^+$ ion is highly hydrated, and this more than compensates for the small size of the anhydrous ion, so that the mobility of Li$^+$ (and Na$^+$) in solution is actually less than that of K$^+$. The mobilities of K$^+$, Rb$^+$ and Cs$^+$ are about equal, since the difference in size of the anhydrous ions is here compensated by the smaller, and diminishing, tendency to hydrate.

Group I (B)

		Cu	Ag	Au
Atomic number		29	47	79
Atomic weight		63·54	107·88	197·0
Atomic structure		2, 8, 18, 1	2, 8, 18, 18, 1	2, 8, 18, 32, 18, 1
Density	(g/cm^3)	8·92	10·50	19·3
Melting point	(°C)	1,083	960	1,063
Boiling point	(°C)	2,580	2,180	2,700
Atomic volume		7·1	10·3	10·2
Atomic radius	Å	1·17	1·34	1·34
Ionic radius	Å	0·96	1·26	(1·37)
Ionic mobility			54	—
Gaseous ionization potential		7·72	7·57	9·22

Here, we have compact atoms of high, and increasing, density; the atom of silver is only a little larger than that of copper, and the gold atom is much the same size as silver. This means that electrons are not very readily removed from the atoms; the metals are not highly electropositive—in fact, decreasingly so with increase of atomic number.

There is not a big decrease of size accompanying the change M → M$^+$, as there is with the alkali metals. While the copper (cupric) ion is highly hydrated, the silver (Ag^+) ion is not; we cannot directly compare the two because of the difference in charge.

Group II (A)

	Be	Mg	Ca	Sr	Ba
Atomic number	4	12	20	38	56
Atomic weight	9·012	24·31	40·08	87·62	137·34
Atomic structure	2, 2	2, 8, 2	2, 8, 8, 2	2, 8, 18, 8, 2	2, 8, 18, 18, 8, 2
Density (g/cm³)	1·86	1·75	1·55	2·6	3·6
Melting point (°C)	1,280	650	850	800	850
Boiling point (°C)	1,500	1,110	1,440	1,370	1,540
Atomic volume	4·85	14·0	26·1	34·0	38·3
Atomic radius (Å)	0·89	1·36	1·74	1·91	1·98
Ionic radius (Å)	0·31	0·65	0·99	1·13	1·35
Ionic mobility	30	55·5	59·8	59·8	64·2
Salts hydrated per cent	80	88	76	78	61
Ionization potential	18·21	15·03	11·87	10·98	9·95

Group II (B)

	Be	Zn	Cd	Hg
Atomic number	4	30	48	80
Atomic weight	9·012	65·37	112·4	200·59
Atomic structure	2, 2	2, 8, 18, 2	2, 8, 18, 18, 2	2, 8, 18, 32, 18, 2
Density (g/cm³)	1·86	7·1	8·6	13·6
Melting point (°C)	1,280	419	321	−39
Boiling point (°C)	1,500	910	770	357
Atomic volume	4·85	9·2	13·0	14·0
Atomic radius (Å)	0·89	1·25	1·41	1·44
Ionic radius (Å)	0·31	0·74	0·97	1·10
Salts hydrated per cent	80	100	85	4·0
Ionization potential (eV)	18·21	17·89	16·84	18·65

The figures illustrate the differences between beryllium and the alkaline earth metals, as well as the similarities between beryllium and zinc. These are mainly due to the fact that the zinc atom is considerably smaller than that of magnesium, although much heavier. This size difference explains the much greater basic strength of magnesium oxide compared with zinc oxide or beryllium oxide, and brings zinc closer to beryllium.

The close relationships of magnesium and the alkaline earths will be apparent, and the relationships are broadly similar to those of the alkali metals. It will be noticed, for instance, that the ionic mobility again rises from magnesium to barium (although less steeply), due to the greater hydration of the lighter ions compared with the heavier ions. Barium salts are less hydrated than calcium salts, and are rarely deliquescent.

Group III (A)

	Al	Sc	Y	La	Ac
Atomic number	13	21	39	57	89
Atomic weight	26·982	44·956	88·905	138·91	227
Atomic structure	2, 8, 3	2, 8, 8, 3	2, 8, 18, 8, 3	2, 8, 18, 18, 8, 3	2, 8, 18, 32, 18, 8, 3
Density (g/cm^3)	2·7	2·5	5·5	6·2	
Melting point (°C)	660	1,420	1,500	920	
Boiling point (°C)	2,500	2,480	3,200	3,300	
Atomic volume	10·0	18·0	16·2	22·4	
Atomic radius (Å)	1·25	1·44	1·62	1·69	
Ionic radius (Å) (3+)	0·50	0·68	0·90	1·06	1·11
(3rd) Ionization potential (eV)	28·44	24·8	20·5	19·17	

In Group IIIB, we get the alternative mode of ionization at (group valency -2), involving an inert pair. This is a marked feature of thallium chemistry.

Group III (B)

	B	Al	Ga	In	Tl
Atomic number	5	13	31	49	81
Atomic weight	10·811	26·982	69·72	114·82	204·37
Atomic structure	2, 3	2, 8, 3	2, 8, 18, 3	2, 8, 18, 18, 3	2, 8, 18, 32, 18, 3
Density (g/cm^3)	2·4	2·7	5·93	7·29	11·85
Melting point (°C)	2,300	660	29·8	156	449
Boiling point (°C)	2,550	2,500	2,070	2,100	1,390
Atomic volume	4·4	10·0	11·8	15·7	17·25
Atomic radius (Å)	0·80	1·25	1·25	1·50	1·55
Ionic radius (Å) (3$^+$)	—	0·50	0·62	0·81	0·95
(1st) Ionization potential (eV)	8·3	5·95	6·0	5·8	6·1
(3rd) Ionization potential	37·92	28·44	30·6	27·9	29·7

Group IV (A)

	Ti	Zr	Hf
Atomic number	22	40	72
Atomic weight	47·90	91·22	178·49
Atomic structure	2, 8, 10, 2	2, 8, 18, 10, 2	2, 8, 18, 32, 10, 2
Density (g/cm^3)	4·50	6·53	13·07
Melting point (°C)	1,725	1,860	2,200
Boiling point (°C)	3,260	4,750	—
Atomic volume	10·64	14·0	13·7
Atomic radius (Å)	1·32	1·45	1·44
(4th) Ionization potential (eV)	43·24	33·8	—

'Ionic radius' has little meaning for these and later group metals, since ions with such high charges are scarce or non-existent. This is because the ionization potentials are large (for the fourth electron).

The close similarity in size of zirconium and hafnium atoms, due to the lanthanide contraction, will be apparent from the figures for atomic volume and atomic radius.

Group IV (B)

	C	Si	Ge	Sn	Pb
Atomic number	6	14	32	50	82
Atomic weight	12·011	28·086	72·59	118·69	207·19
Atomic structure	2, 4	2, 8, 4	2, 8, 18, 4	2, 8, 18, 18, 4	2, 8, 18, 32, 18, 4
Density (g/cm^3)	3·5[1] 2·25[2]	2·49	5·36	7·3	11·3
Melting point (°C)	3,500	1,400	960	230	327
Boiling point (°C)	—	2,300	2,700	2,360	1,750
Atomic volume	3·4[1] 5·3[2]	11·4	13·6	16·2	18·3
Atomic radius (Å)	0·77	1·17	1·22	1·41	1·54
Ionic radius (M^{2+})					1·32
(2nd) Ionization potential (eV)	—	—	15·86	14·5	14·96

[1] diamond.   [2] graphite (the other elements also show allotropy).

In subgroup IVB, we get a continuation of 'inert pair' behaviour, particularly for tin and lead.

Group V (A)

		V	Nb	Ta
Atomic number		23	41	73
Atomic weight		50·942	92·906	180·948
Atomic structure		2, 8, 11, 2	2, 8, 18, 11, 2	2, 8, 18, 32, 11, 2
Density	(g/cm^3)	5·96	8·4	16·6
Melting point	(°C)	1,700	2,400	2,850
Atomic volume		8·4	10·8	10·9
Atomic radius	(Å)	1·22	1·34	1·34

The boiling points are very high and not very accurately determined. As in Group IVA, the second and third elements are very similar in the size of their atoms and in chemical properties; this consequence of the lanthanide contraction is a general property of transition elements.

Group V (B)

		N	P	As	Sb	Bi
Atomic number		7	15	33	51	83
Atomic weight		14·0067	30·974	74·922	121·75	208·98
Atomic structure		2, 5	2, 8, 5	2, 8, 18, 5	2, 8, 18, 5	2, 8, 32, 18, 5
Density	(g/cm^3)	1·03	1·83	5·7	6·6	9·8
Melting point	(°C)	−210	16·96	(subl.)	630	270
Boiling point	(°C)	−196	287	616	1,440	1,420
Atomic volume		13·65	16·96	13·3	18·5	21·3
Atomic radius	(Å)	0·74	1·1	1·21	1·41	1·52
Ionic radius	(M^{3+})	—	—	—	0·90	1·20
Electronegativity (Pauling)		3·0	2·1	2·0	1·8	1·8

The electronegativity scale was worked out first by Pauling on a somewhat arbitrary basis, and compares the relative attractions of the elements for electrons (see p. 89). This series of elements is decreasingly electronegative, or increasingly electropositive, with increase in atomic number. The elements arsenic, antimony and bismuth show valency three, due to the inert pair (as in some other B groups).

Group VI (A)

		Cr	Mo	W
Atomic number		24	42	74
Atomic weight		51·996	95·94	183·85
Atomic structure		2, 8, 13, 1	2, 8, 18, 13, 1	2, 8, 18, 32, 12, 2
Density	(g/cm^3)	7·1	10·4	19·3
Melting point	(°C)	1920	2600	3400
Atomic volume		7·3	9·4	9·5
Atomic radius	(Å)	1·17	1·29	1·30

Again the boiling points are very high; melting and boiling points reach maxima in each period in the middle transition groups (V and VI). The atoms are compact, of high density and small size.

Group VI (B)

	O	S	Se	Te	Po
Atomic number	8	16	34	52	84
Atomic weight	15·9994	32·064	78·96	127·60	210
Atomic structure	2, 6	2, 8, 6	2, 8, 18, 6	2, 8, 18, 18, 6	2, 8, 18, 32, 18, 6
Density (g/cm^3)	1·27	2·06	4·8	6·2	9·5
Melting point (°C)	−219	115	217	450	250
Boiling point (°C)	−183	445	685	1,400	960
Atomic volume	12·6	15·6	16·5	20·2	22·2
Atomic radius (Å)	0·74	1·04	1·17	1·37	1·64
Ionic radius (M^{2-})	1·40	1·85	1·98	2·21	
Electronegativity	3·5	2·5	2·4	2·1	

The electronegativity of these elements is greater than that of the Group VB elements, and again falls off with increasing atomic number. The elements form simple anions, which are considerably larger than the covalent, neutral atoms.

Group VII (A)

	Mn	Tc	Re
Atomic number	25	43	75
Atomic weight	54·938	99	186·2
Atomic structure	2, 8, 13, 2	2, 8, 18, 13, 2	2, 8, 18, 32, 13, 2
Density (g/cm^3)	7·4	11·5	20·5
Melting point (°C)	1260	—	3170
Atomic volume	7·4	8·6	8·8
Atomic radius (Å)	1·17	—	1·28

THE PERIODIC TABLE

Group VII (B)

		F	Cl	Br	I
Atomic number		9	17	35	53
Atomic weight		18·998	35·453	79·909	126·904
Atomic structure		2, 7	2, 8, 7	2, 8, 18, 7	2, 8, 18, 18, 7
Density	(g/cm³)	1·11	1·56	3·12	4·93
Melting point	(°C)	−233	−102	−7	113
Boiling point	(°C)	−188	−35	59	183
Atomic volume		17·1	18·7	23·5	25·7
Atomic radius	(Å)	0·72	0·99	1·14	1·33
Ionic radius	(M⁻)	1·36	1·81	1·95	2·16
Electronegativity		4·0	3·0	2·8	2·5
Ionic mobility		46·6	65·5	67·6	66·5

These figures show that similar relationships exist in Group VIIB and Group VIB, but the halogens are rather more electronegative.

Group VIII

		Fe	Co	Ni
Atomic number		26	27	28
Atomic weight		55·847	58·933	58·71
Atomic structure		2, 8, 14, 2	2, 8, 15, 2	2, 8, 16, 2
Density	(g/cm³)	7·9	8·7	8·9
Melting point	(°C)	1,535	1,480	1,450
Boiling point	(°C)	2,890	2,880	2,840
Atomic volume		7·1	6·7	6·7
Atomic radius	(Å)	1·16	1·16	1·15
Ionic radius (M^{2+})	(Å)	0·83	0·82	0·78

It will be seen that these elements closely resemble each other in physical properties.

TABLES OF PHYSICAL CONSTANTS

	Ru	Os	Rh	Ir	Pd	Pt
Atomic number	44	76	45	77	46	78
Atomic weight	101·07	190·2	102·91	194·2	106·4	195·09
Atomic structure	2, 8, 18, 15, 1	2, 8, 18, 32 14·2	2, 8, 18, 16, 1	2, 8, 18, 32, 17	2, 8, 18, 18	2, 8, 18, 32, 17, 1
Density (g/cm^3)	12·2	22·5	12·4	22·4	11·9	21·4
Melting point (°C)	2,500	2,700	1,970	2,450	1,560	1,770
Atomic volume	8·6	8·5	8·8	8·6	9·0	9·1
Atomic radius (Å)	1·24	1·26	1·25	1·26	1·28	1·29

These six elements again resemble each other in physical properties; in particular their sizes are all very similar. These very small atoms are most reluctant to lose electrons, and their compounds are covalent.

Group 0

	He	Ne	Ar	Kr	Xe	Rn
Atomic number	2	10	18	36	54	86
Atomic weight	4·003	20·183	39·944	83·80	131·30	222
Atomic structure	2	2(2,6)	2.8(2,6)	2.8.18, (2,6)	2.8.18.18, (2,6)	2.8.18.32.18, (2,6)
Melting point (°K)		24	84	116	161	202
Boiling point (°K)	4·2	27	87	121	164	211
Atomic radius (°A)	1·2	1·6	1·9	2·0	2·2	—
(1st) Ionization potential (eV)	24·5	21·5	15·7	13·9	12·1	10·7

BIBLIOGRAPHY

The author acknowledges his indebtedness to the authors of the following books in particular.

The Chemical Elements and their Compounds, N. V. Sidgwick, University Press, Oxford, 1950

Modern Aspects of Inorganic Chemistry, H. J. Emeleus and J. S. Anderson, Routledge & Kegan Paul, London, 1960

The Nature of the Chemical Bond, L. Pauling, University Press, Oxford, 1940

Chemistry of the Lanthanons, R. C. Vickery, Butterworths, London, 1953

Valency and Molecular Structure, E. Cartmell and G. W. A. Fowles, Butterworths, London, 1956

Sourcebook on Atomic Energy, S. Glasstone, Macmillan, London, 1952

The Chemistry of the Actinide Elements, J. J. Katz and G. T. Seaborg, Methuen, London, 1957

Inorganic Chemistry, R. B. Heslop and P. L. Robinson, Elsevier, Amsterdam, 1960

Structural Principles in Inorganic Compounds, W. E. Addison, Longmans, London, 1961

INDEX

Abundance of elements, 97
Actinium, 50, 80
Actinons, 10, 80, 96
Alkali metals, 13, 17, 104
Alkali metal compounds, 18
Alkaline earth metals, 21, 106
    compounds, 22
Allotropy, 28, 31, 37
Aluminium, 12, 24, 108
Americium, 85
Antimony, 31, 110
Argon, 47
Arsenic, 31, 110
Astatine, 43
Atomic
    radius, 2, 5, 104
    volume, 1, 3, 17, 104

Barium, 21, 106
Beryllium, 11, 21, 106
Bismuth, 15, 31, 110
Boron, 12, 24, 108
Bromine, 43, 112

Cadmium, 76, 106
Caesium, 17, 104
Calcium, 21, 106
Carbon, 28, 109
CARTLEDGE, 13
Catalytic activity, 94
Chlorides, 103
Chlorine, 15, 43, 112
Chromium, 59, 110
Clathrate compounds, 48
Cobalt, 66, 112
Coinage metals, 72
Coloured ions, 94
Complex formation, 23, 74, 78, 88, 95
Conductivity, 103

Copper, 72, 105
Covalency maximum, 12, 29

Density, 1
Diagonal relationship, 6, 11, 76
Dielectric constant, 38
Dipole moment, 44
Displacement law, 100

Electrode potential, 78
Electron affinity, 6, 14
Electron-deficient compounds, 24
Electron orbitals, 7, 92
Electronegativity, 89
Electropositive series, 86
Electropositivity, 86
Emanation, 49
Energy levels, 8, 83, 93
Even number rule, 101

FAJANS, 11, 100
FAJANS' rules, 11, 13, 19, 86, 94
Ferrous metals, 66, 112
Flame spectra, 19
Fluorine, 43, 112

Gallium, 24, 108
Germanium, 28, 109
Gold, 72, 105
GOLDSCHMIDT, 97
Grignard reagents, 22

Hafnium, 52, 108
Halogens, 43, 112
Helium, 47
Heteropoly acids, 35, 60

INDEX

Hydration of salts, 12, 19, 22, 87, 105
Hydrogen, 13
Hydrolysis in solution, 12, 26, 88

Indium, 24, 108
Inert gases, 47
Inert pair, 15, 26, 33, 40
Inner transition elements, 10, 80
Iodine, 43, 112
Ion exchange, 50, 53, 81
Ionic
 mobility, 19, 105
 potential, 13
 radius, 5, 105
Ionization potential, 6, 16, 81
Iridium, 66, 113
Iron, 66, 112
Isomorphism, 99
Isopoly acids, 35, 60

Krypton, 47

Lanthanide contraction, 53, 59, 81
Lanthanons, 80
Lanthanum, 50, 80, 107
Lead, 15, 28, 109
Lithium, 17, 104
Lutetium, 80

Magnesium, 12, 21, 106
Manganese, 62, 111
Melting point, 1, 103, 104
MENDELEEFF, 3
Mercury, 76, 106
MEYER, LOTHAR, 3
Molybdenum, 59, 110

Neon, 47
Neptunium, 83
Neutron absorption, 83, 102
Neutron–proton ratio, 101
Nickel, 66, 112
Niobium, 55, 109
Nitrogen, 31, 110

Nobelium, 83
Noble gases, 47
Noble metals, 72, 105
 compounds, 73

Occurrence of elements, 98
Orbitals, 7, 92
Osmium, 66, 113
Oxygen, 37, 111

Palladium, 66, 113
Paramagnetism, 94
α-particle, 100
β-particle, 100
PAULING, 89
Periodicity, 2
Phosphorus, 31, 110
Photo-emission, 18
Platinum, 66, 113
Plutonium, 83
Polonium, 37, 111
Potassium, 17, 104

Radioactivity, 100
Radium, 21, 106
Radon, 47
RAYLEIGH, 47
Rhenium, 62, 111
Rhodium, 66, 113
Rubidium, 17, 104
RUSSELL, 100
Ruthenium, 66, 113

Salts of oxyacids, 87
Scandium, 50, 107
SEABORG, 83
Selenium, 37, 111
SIDGWICK, 66
Silicon, 28, 109
Silver, 72, 105
SODDY, 100
Sodium, 17, 104
Solubility, 98
Strontium, 21, 106
Sulphur, 37, 111

Tables of physical constants, 104
Tantalum, 55, 109
Technetium, 62, 111

INDEX

Tellurium, 37, 111
Thallium, 15, 24, 104
Tin, 15, 28, 109
Titanium, 52, 108
Transition properties, 9, 65, 85, 91
Transuranic elements, 83
Tungsten, 59, 110

Valency, 4
Vanadium, 55, 109

Variable valency, 93
Volatility, 98

Water, 38

Xenon, 47

Yttrium, 50, 107

Zinc, 76, 106
Zirconium, 52, 108